WORKING WITH CONGRESS

A Practical Guide for Scientists and Engineers

WORKING WITH CONGRESS

A Practical Guide for Scientists and Engineers

Second Edition

WILLIAM G. WELLS, JR.

AMERICAN ASSOCIATION FOR THE
ADVANCEMENT OF SCIENCE

Library of Congress Catalog-in-Publication Data

Wells, William G., Jr.
Working with Congress: a practical guide for scientists and engineers /
prepared by William G. Wells, Jr. for the American Association for the
Advancement of Science and the Carnegie Commission on Science,
Technology, and the Government.
    p.   cm.
    Includes bibliographic references.
    ISBN 0-87168-581-7
    1. Scientists-United States-Handbooks, manuals, etc.
2. Engineers—United States-Handbooks, manuals, etc.   3. Science and
state—United States—Handbooks, manuals, etc.   4. Engineering and
state—United States—Handbooks, manuals, etc.   5. United States.
Congress—United States—Handbooks, manuals, etc.   I. Title.
Q149.U5W46   1996
338.97306—dc20   92-35313
                                                              CIP

The AAAS Board of Directors, in accordance with Association policy, has
approved the publication of this work as a contribution to the
understanding of an important area. Any interpretations and conclusions
are that of the author and do not necessarily represent views of the Board
or the Council of the Association.

Design by Peggy Friedlander

AAAS Publication: 96-2S
International Standard Book Number: 0-87168-581-7

Printed in the United States of America

Printed on recycled paper

TABLE OF CONTENTS

PREFACE

Both scientists and elected officials share a mandate to work in the interest of the nation. Although one can easily point to their differences, they have much in common, and important forces bring them together. Science works for society, and society, in turn, depends on and nurtures science.

Changes in Congress brought about first by calls to reduce the deficit in the early 1990s and strengthened by the "Republican Revolution" of 1994 seemed to warrant a second look at *Working with Congress: A Practical Guide for Scientists and Engineers*. More and more groups, both here in Washington and across the country, are calling upon individual scientists and engineers to build stronger relationships with their senators and representatives. This revised second edition offers scientists and engineers clear, concise advice on how to communicate with lawmakers and their staffs. As events of recent years have demonstrated, Congress, as the source of the nation's laws and the appropriator of federal funds, exerts considerable influence over the future directions of science and technology. This book can help scientists and engineers develop a basic understanding of congressional operations in order to better serve their interests and the interests of the nation.

The information and advice in this book comes from extensive discussions with members of Congress and their staffs, from a written survey done for the original version of the book and a second survey done for this edition, and from my own personal experience and the experience of many others in working with Congress. This edition of the *Working with Congress* incorporates information about communicating with members via electronic mail and includes a new section in "Appendix C: Sources of Information about Congress" that describes World Wide Web sites on the Internet that pertain to Congress. The list of professional societies and other organizations dealing with science and technology policy (Appendix B) has also been updated and enlarged. The Seventeen Cardinal Rules for working with Congress still apply, despite the dramatic changes of the last five years. They are discussed in chapter four and expanded on in chapter five.

Many people contributed to this volume. The members and staff who took time to answer the questionnaires or speak with me deserve special thanks. I would like to thank Elaine Furlow for her initial review of the original manuscript and Gretchen Richter for coordinating the second survey of members and staff and for her work on the appendices. Finally, I am deeply grateful to the staff of the AAAS Directorate for Science and Policy Programs and its Center for Science, Technology, and Congress, specifically, Bonnie Bisol Cassidy and Al Teich for their collaboration during the revision process and Celia McEnaney for her oversight of the production process; without them this new edition would not have been published.

FOREWORD

The 1990s are a time of great change in Congress: Republicans have taken control of the House for the first time in 40 years and regained control in the Senate after ten years in the minority; the number of junior members reached an all time high in 1995; the number of committees and subcommittees has been reduced, as have staff levels; and a record number of representatives have switched parties. In the midst of all these changes, Congress has undertaken a serious effort to balance the budget and reduce the size of the federal government. This near-constant state of flux makes it more imperative than ever that scientists and engineers communicate effectively with Congress about the importance of their work.

The congressional agenda is increasingly dominated by issues involving science and technology, but relatively few members of Congress or congressional staff have any training or background in science or engineering. Responding effectively to complex scientific and technical challenges such as economic competitiveness; global environmental change; the AIDS epidemic; the ethical, legal, and social issues of genetics research; energy uncertainties; and national security requires that Congress have access to the best scientific and technical advice. However, although many scientists and engineers have contact with the executive branch through grant support, participation on advisory committees, and other mechanisms, only a handful have much experience dealing with Congress. Observers both inside and outside of Congress have noted that scientists and engineers have not been as effective as they could be in their interactions with lawmakers and have suggested that improved communication could benefit both the research community and the quality of national policymaking.

Since the beginning of 1995, many groups and coalitions have formed (or reenergized) with the aim of protecting the federal investment in science and technology. These groups have been encouraging members of the scientifc and engineering communities to establish long-term relationships with members of Congress. The days of steadily increasing research budgets are over and are not likely to return in the forseeable future. However, members of Congress, especially the newer members, resent what they perceive as an assumption that research funding is an "entitlement." Members of the science and technology community must begin to take responsibility for explaining the significance of their work and demonstrate the return on the federal investment in research.

The American Association for the Advancement of Science (AAAS) has long been interested in improving the relationships at the intersections of science, technology, and government and in better integrating information on science and technology into policymaking. Founded in 1848, AAAS is the largest general scientific organization in the world, with over 143,000 scientists, engineers, educators, policymakers, and interested citizens among its individual members. In 1994, AAAS established the Center for Science, Technology, and Congress, funded by a grant from the Carnegie Corporation of New York, to provide timely, objective information to Congress about issues in science and technology and to help enhance communication between scientists and engineers and policymakers. This second edition of *Working with Congress* is one part of the integrated program of activities undertaken by the Center to improve the understanding and use of scientific and technological information on Capitol Hill. The book describes the constitutional basis of Congress, its culture and traditions, its power structure and organization, and its principal activities in a way that is intended to help scientists and engineers understand how Congress works and how best to interact with members and staff. It is a practical manual to assist scientists and engineers in working with Congress—whether it be through personal visits, telephone, fax, and e-mail interactions, correspondence, or participation in hearings.

If scientists and engineers expect members of Congress to understand and appreciate the significance of scientific and technological endeavors, it is critical that they make an effort to understand and work with Congress. Hopefully, this book will encourage you, the reader, to become more involved in congressional activities. Congressional involvement can be highly rewarding—there is satisfaction to be gained from seeing one's ideas influence policy and one's expertise used to benefit the nation.

Bonnie Bisol Cassidy
Assistant Director
AAAS Center for Science, Technology, and Congress

Albert H. Teich
Director
AAAS Directorate for Science and Policy Programs

1 WHY WORK WITH CONGRESS?

Representative George E. Brown, Jr., a longtime congressman from California and ranking Democrat on the House Science Committee, remembers a veterinarian from his district who visits him once a year. "Now there are not many bills that come through Congress that involve a veterinary issue," recalled Brown, "But when such an issue or question does come up, I pay attention. If I need a veterinarian's opinion, I remember him."

Many people call their representative or senators only when they need something. But Representative Brown's story points out the importance of staying in touch with your representative or senators on a more regular basis, often enough to become known as a familiar, trustworthy source of scientific or technical opinion. "If you lay groundwork on your issue in a clear, concise way, you set the stage for a member to pay attention when you need him to," advises one staffer.

The hard part for scientists and engineers may come in knowing how to work with Congress in a comfortable, effective way. Just as technical people may use a different language and different working style than do business people, members and staff in Congress use a language and working style that may, to onlookers, seem hard to understand. That's where this guide comes in. It is a behind-the-scenes, between-the-lines look at how Congress works and how a scientist or engineer can interact with Congress effectively.

The interaction can take place in Washington or in one's own state or district. Wherever it takes place, and whether it's in person or by letter, the key is talking in a language the people in Congress understand and value. By framing what you do in a context that lawmakers can relate to, you increase the chances that your message will be heard and will get the response you desire.

This guide is a primer for how to do just that. The payoff? When a member of Congress hears and responds to your views on important issues, that exchange benefits the making of public policy, the discipline or organization you represent, and sometimes even you yourself.

Congressional activities take time, and in today's highly competitive environment, most scientists and engineers have barely enough time to do their research, pursue funding to keep it going, and fulfill their teaching, administrative, and professional responsibilities. Furthermore, many technically trained people are put off by the rhetoric and seeming irrationality of much of what goes on in Congress. They are more comfortable in the laboratory or out in the field than in the highly charged political atmosphere of Capitol Hill. Then, too, they are skeptical of having any meaningful impact when they are in competition with so many professional lobbyists and more powerful interests.

Most individuals have no such reservations about dealing with the executive branch of the federal government. In fact, it is a rare scientist or engineer who does not have some connection with one or more federal agencies. Academics might be involved with the National Science Foundation or the National Institutes of Health, either as a recipient of grant, contract, or fellowship support; as a member of a review or advisory panel; or as a colleague or collaborator with a government researcher. Scientists, engineers, or researchers in private or nonprofit organizations might collaborate with the National Aeronautics and Space Administration, the Environmental Protection Agency, or the departments of Defense, Energy, and Agriculture. With these growing linkages between science, technology, and government, scientists and

" IN LAYMAN'S TERMS? I'M AFRAID I DON'T KNOW ANY
LAYMAN'S TERMS. "

engineers have important reasons for being in direct, frequent contact with executive branch agencies. The interweaving of science, technology, and government has become inevitable and increasingly important to both sides. But Congress is different. In general, Congress does not provide funds directly to individual institutions or research projects. Instead, it sets broad policy directions or provides money for programs which are carried out by executive branch agencies. Thus, few scientists and engineers may feel they need to work with Congress, and relatively few do so regularly.

But the scientific community cannot afford to regard working with Congress as a casual matter—it is essential today and will be even more vital in the future. This guide gives you the tools and understanding to increase your effectiveness with those in Congress and can make your experience a positive and productive one.

3

Working with Congress serves the public and national interest, serves the interests of science and engineering, and also serves one's own personal or institutional self-interest. Scientists and engineers can serve the public and national interest by providing advice, information, and opinions based on both specific scientific and technical expertise and general understanding of science and technology. The applications of science and technology to policy—for example, the scientific understanding of global climate change—are much broader than the questions that involve policy for science and technology, such as the funding levels for research endeavors in various agencies. Scientists and engineers can contribute in both arenas. They can offer views and information relating to government programs, regulations, and other policy issues based on their own research and expertise, or, in some instances, on the basis of their general scientific and technical knowledge. Of necessity, such views reflect judgments that go beyond purely scientific considerations and into the realm of values. Politicians are accustomed to dealing with advice in these terms, but scientists and engineers who offer it should still strive to make clear when their ideas are based on generally accepted data and interpretations and when they are personal opinions or judgments.

Helping to make good policy on issues involving science and technology is a worthwhile pursuit for scientists. But scientists and engineers in all fields have an obvious interest in the welfare of the research enterprise, as well as their discipline and research area and their levels of federal support. There is no reason to be shy about pursuing this kind of indirect self-interest by acting as an advocate for science and engineering or one's own field. Carried out in an appropriate, responsible manner, such advocacy is normal, respectable behavior in the political environment.

In terms of direct self-interest, virtually all scientists and engineers in universities and government and a great many in industry and other institutions are profoundly affected by federal policies and budgets. A researcher might support increased funding for an area of research because support for his or her own institution, department, or laboratory would stand to benefit from larger appropriations. Political leaders expect citizens to pursue their own interests and scientists and engineers should not hesitate to do so.

Should scientists and engineers become more involved in working with Congress? The alternative is to leave science and technology policymaking in the hands of other groups and interests. Congress will make decisions on support for science and engineering research and on other science and technology policy issues whether scientists and engineers choose to become involved or not. To ignore Congress or to remain aloof is to forego the chance to influence policy and to abdicate one's responsibility to the science and engineering communities—and to the nation.

THE CONGRESSIONAL ARENA

2

This chapter provides an overview of Congress, including its constitutional setting, distinguishing features, and relationships with the other branches of government. The focus is on aspects that should interest those who wish to work with Congress but who are not professional scholars or analysts of Congress.

CONSTITUTIONAL SETTING OF CONGRESS

The United States Constitution is a product of consummately crafted compromises arising from long months of negotiations, posturing, threats, and patient statesmanship at the Constitutional Convention of 1787. Take an hour to browse through it. The Constitution is a remarkable document that represents a stunning intellectual and political achievement. A prime example of "less is more," it is founded on a handful of basic principles: limited government, separation of powers, checks and balances, federalism, and popular control of government.[1]

These principles continue to guide lawmaking and the operations of government today, despite the enormous changes that have transformed the world and the role of government in American society over the past two hundred years. Consider for a moment the Constitution's durability and continuing relevance in this world of change, and how these principles have served the nation.

LIMITED GOVERNMENT

The nation builders who crafted the Constitution sought to establish a strong and effective national government, but at the same time they wanted to protect personal and property rights by avoiding the concentration of too much power in that government. The limitations were achieved by dividing power among the three branches of government and among the national government, the states, and the citizens. And the first ten amendments to the Constitution, the Bill of Rights, are brilliantly designed to ensure the rights and liberties of the citizens and to protect them against an overreaching government.

SEPARATION OF POWERS

The framers of the Constitution wanted to guard against a monopoly of governing power, so they established three independent branches of government. This limits and constrains the authority of any one branch, but it also ensures cooperation among the three branches: such cooperation is necessary to make things work in practice. Yet the framers of the Constitution clearly had a distinct bias toward representative assemblies. Article I of the Constitution shows a sharp, central focus on the powers of Congress. The framers of the Constitution identified the powers of Congress at length, named Congress as the first branch of government, and, in case they missed something, granted Congress a wide range of explicit and implied powers via the so-called elastic clause (Article I, Section 8).

> Article I, Section 1. "All legislative Powers herein granted shall be vested in a Congress of the United States, which shall consist of a Senate and House of Representatives."

Article I, Section 8 [Following a long list of congressional powers, this section adds] "To make all Laws which shall be necessary and proper for carrying into Execution the foregoing Powers, and all other Powers vested by this Constitution in the Government of the United States, or in any Department or Officer thereof."

In contrast, Article II describes the framework of the executive branch and its duties in short, vague terms. The Constitutional Convention was sharply divided on the issue of executive power; ultimately they settled on a compromise to get an agreement. Thus, the Constitution reflects the struggle between two conceptions of executive power: on the one hand, it ought always to be subordinate to the supreme legislative power; on the other hand, it ought to be, within limits, autonomous and self-directing. The practical effect was to plant the seeds of implied presidential power and the long-term development of the presidency as "what the President thinks it is," at least within the limits set by the system of checks and balances.

Article III, which establishes the judiciary, is even leaner than Article II. Here again, Congress has a role, being given the power to establish new judicial positions and create new courts below the Supreme Court. The sparseness of the constitutional language enabled the fourth chief justice, John Marshall, to establish the principle of judicial review, providing the federal courts with the power to rule on the constitutionality of specific legislative and executive decisions. Marshall's role illustrates an important dimension of the flexibility of the Constitution and the American political system. Within limits, the nation's affairs can be tremendously affected by a strong and determined individual whether in Congress, the presidency, or the judiciary.

CHECKS AND BALANCES

Anticipating that strong individuals in each branch might seek to expand their own power at the expense of the other branches, the framers actually crafted what amounts to an open invitation to conflict and struggles for power, especially between Congress and the President. To contain this conflict, the framers devised an intricate system of checks and balances.

Examples include the ability of the President to veto laws passed by Congress; the internal constraints within Congress arising from its bicameral nature; and the requirement that budgets, treaties, and high-level appointments proposed by the President must be approved by Congress. In addition, many decisions and actions of Congress and the President are subject to review by the federal judiciary. Finally, judicial decisions—including those of a constitutional nature—may be overturned by Congress and the President working in concert. Such checks and balances have a dual effect: they introduce the potential for conflict, but they ultimately encourage cooperation, negotiation, compromise, and accommodation among the branches.

FEDERALISM

Just as the three branches of the national government check each other, the Constitution includes another dimension of checks and balances and division of power between the state and federal governments. Powers not granted to the national government by the Constitution remain with the states and with the people.

POPULAR CONTROL OF GOVERNMENT

The framers of the Constitution intended the House of Representatives to be the most representative part of the national government. Representatives are elected directly by the people for two-year terms to ensure that the people's will is continuously reflected within the government. In practice, for many members of the House this means campaigning non-stop—visiting their districts, making speeches, serving constituents, and raising campaign money—as well as attending to the lawmaking responsibilities of a member of Congress.

Originally, the Senate was designed to be one step removed from popular voting in order to moderate the popular will of the House. State legislatures were given the responsibility of electing senators. In 1913, this process was swept aside with the enactment of the Seventeenth Amendment, which provides for direct election of senators. While senators serve six-year terms, and thus do not face elections as frequently as House members, they too have become intensely responsive to constituents. Members of Congress are attuned to the power of the people to "turn the rascals out" if they so choose. From time to time, that is exactly what the people do.

HOW MEMBERS VIEW CONGRESS AND THEMSELVES

Members of Congress, along with the President and the Vice President, are acutely aware of belonging to a highly select group that shares an almost mystical bond: they were elected to national office by the citizens of the United States. They never forget this and do not want others to forget it either. Membership in this very special community separates them constitutionally and in their heart of hearts from appointed officials—including even Supreme Court justices and cabinet secretaries—and from staff members, however close such individuals might be professionally or personally. It is a bond that transcends partisan concerns and branches of government.

However green and untutored new members of Congress may be upon arrival in Washington, they soon absorb the implications of their initiation into this select society. Although members of Congress can be as open as the average citizen to flattery and courting by the President and as impressed as anyone by a personal call from the Oval Office, they nevertheless see themselves as constitutionally coequal with the President in a collective sense. More than a few Presidents (and their staffs) have had to learn this the hard way. When a President attempts to bully or threaten a member, particularly in private, the member will more often than not respond in kind. Those Presidents who have been most effective in getting their programs passed have tended to treat members of Congress as equals.

It is important to understand that no matter how powerful or high level a congressional or executive staff person may be, it is perilous for them to presume equality with a senator or representative. Sometimes senior individuals from industry or academia who have been appointed to upper level executive branch positions have an encounter with a member of Congress that clearly places them in a status inferior to even a junior member. Even White House staff members, as powerful as many of them believe themselves to be, must tread carefully in dealing with a member of Congress—including those of their own party. Obviously, there are some exceptions to this rule, but the general point remains valid.

As a constitutional principle, the separation of powers is central to our political system. It is built into the very fiber of individual members of Congress and the institution collectively. It is a concept that often outweighs party bonds. While a President usually can count on support from members of Congress from his own party, such support is not automatic. Indeed, support will be withheld or given only very grudgingly if it appears that congressional prerogatives are being slighted. If they are seen as being trampled upon, the President can almost certainly expect opposition even from members of his own party. Such behavior is not simply a matter of personal pique; it rests on a deep sense of constitutional integrity.

Finally, a less-than-attractive aspect of membership in this select society sometimes emerges in the form of arrogance and domineering personal behavior. The Hill is organized to cater to the commands and wishes of members of Congress. Perks abound in great variety (although these may be somewhat restricted with the gift bans and other reform efforts of the 104th Congress); staff of all types (personal, support, committee) work furiously to satisfy member requests, the Capitol police stop traffic for members to cross the street, and every facet of congressional operations is geared to support the notion of constitutional uniqueness and privilege of members. It takes a strong personality to resist the blandishments of life on Capitol Hill.

CONGRESSIONAL DUALITY

Representative assemblies contain an inherent tension between lawmaking and representation. Roger H. Davidson and Walter J. Oleszek start their excellent analysis, *Congress and Its Members*, with two quotes that express this central thesis about Congress: its fundamental duality.

> A good government implies two things: first, fidelity to the object of government, which is the happiness of the people, secondly, a knowledge of the means by which that object can be best obtained. – *The Federalist*, No. 62 (1788)

> Legislatures are really two objects: a collectivity and an institution. As a collectivity, individual representatives act as receptors, reflecting the needs and wants of constituents. As an institution, the Legislature has to make laws, arriving at some conclusions about what ought to be done about public problems. – Charles O. Jones, "From the Suffrage of the People: An Essay of Support and Worry for Legislatures" (1974)

Davidson and Oleszek use these two statements to capture the essence of the dualism of Congress or in their words, the concept of "two Congresses."[2] Political scientist Giovanni Sartori has distinguished between the two functions of Congress by describing them as the "function of functioning" and the "function of mirroring."[3]

According to scholars, observers of Congress, and members themselves, one of the "two Congresses" is a lawmaking institution. This is the Congress of most textbooks, the media, and especially the daily network television news. It can be seen as a collegial body, performing constitutional duties and handling legislative issues. This Congress serves as an arena of political combat where many of the forces of American, and increasingly international, politics converge. These forces are at work within an intricate network of congressional structures, procedures, and personalities that sets the rules and organizes the legislative struggles.[4]

But there is a second Congress, the representative assemblage of 540 senators, representatives, and delegates.[5] Despite differing ages, diverse backgrounds, and varying routes taken to office, these men and women share a common experience. Their political lives and electoral prospects depend upon the support, goodwill, and perceptions of voters in their districts and states. This leads to arrangements that are built on service to constituents in all its forms. From this concept arises an important truth that one must grasp to achieve a real understanding of Congress. By no means does all congressional activity take place on Capitol Hill.[6] Much of it goes on in the cities, towns, and farms far removed from Washington.

IMPLICATIONS OF CONGRESSIONAL DUALITY

The duality of Congress is manifested first in the schedules of members and the enormous time pressures on those schedules. A member's institutional and individual duties are always there, as is the tension between these dual dimensions of the member's schedule. Virtually every member of Congress would endorse Davidson and Oleszek's observation: "Senators and Representatives suffer from a lack of time to accomplish what is expected of them. No problem vexes Members more than that of juggling constituency and legislative tasks."[7] There is very little time to read and study, although members are required to make major policy decisions. A comprehensive study by the House Commission on Administrative Review (the Obey Commission) revealed that the average day for representative exceeded eleven hours and that only a very small part of the day was available for reading or thinking time. By far, the largest components of time are devoted to meetings in the office and in committees.[8]

CONSTITUENCY BUSINESS—DEMANDS AND OPPORTUNITIES

The average representative spends about 120 days a year in the district; the average senator 80. Even in Washington, some evidence suggests that members spend less than half of their time on lawmaking and related duties and the balance on constituency activities.[9] The pull of constituency business is relentless. Three anecdotes point to the special nature of the constituent-member relationship.

- At midnight, a midwestern representative, accompanied by an aide, wearily dragged himself into a roadside diner for a cup of coffee before heading to the local motel. The day had started at 6 a.m. with a breakfast meeting at a plant thirty miles away. Events had been back-to-back all day. As the coffee arrived, a voice came from a man plunking himself onto the stool next to the Representative. "Hello, Congressman. I'm glad to run into you. I wanted to talk to you about a West Point appointment for my nephew." In the member's own words: "The last thing in the world I wanted to do was talk about West Point; I was so damn tired I couldn't think, but I had to turn on the smile and ask my assistant to get out his notebook to take down the information. In many ways this is a 24-hour-a-day, seven-day-a-week job, but I always remember Harry Truman's advice: 'If you can't stand the heat, get out of the kitchen.'"

- A western senator had gone into a bar on a hot, dusty afternoon to cool off with a beer. Barely into his first sip, he noticed a woman getting up from a table and approaching him with a determined look. After placing herself squarely to his right with feet wide apart, she let go with a stream of profanity about how he was not handling his job, mentioning several local issues in particular. Turning on her heel, she returned to her table and friends, leaving the senator to his beer. A few minutes later, she got up again and came over. But this time she said, "Excuse me, Senator, but I forgot to ask for your autograph for my son."

■ A midwestern Democratic representative explains the organization of his upcoming district political campaign: "Some of my most ardent supporters and tireless volunteer workers are Republican women whose military sons have been assisted through our constituent service. But we never ask their politics when they need help; I serve all of my constituents. If they decide to reciprocate later, that is up to them." He adds that there is a lot more than constituent service in a campaign but concludes, "A lot of my colleagues have seen the light."

The "light" originates not only in the experiences of members, but also in the research of congressional scholars like Richard F. Fenno, Jr., who spent nearly a decade visiting and traveling with members in their districts to gain a better understanding of that "other Congress" away from Washington. Fenno, in commenting on a particular representative, said: "For a congressman facing the prospect of close elections in a district opposed to him philosophically, the appeal of a constituency service emphasis would have to be strong. Constituency service is totally nonpartisan and nonideological. As an electoral increment, it is an unadulterated plus." Fenno concludes that constituent service is especially appealing to a representative of a highly heterogeneous district or one that is bewildering from the standpoint of issue politics.[10]

PUBLIC PERCEPTIONS OF INDIVIDUALS AND INSTITUTIONS

The duality of Congress goes a long way toward explaining a well-known paradox in public perceptions of Congress. The lawmaking (institutional) Congress is often seen as messy and incompetent and is held in low esteem by the public. The institutional Congress seems to be evaluated by citizens on the basis of their overall attitudes about politics, government policies, and the state of the nation.[11] Also, disapproval of the misbehavior of a small minority of its members tends to spill over and color the public's views on the entire institution.

In contrast, citizens measure the performance of their members differently from that of the institution as a whole. They seem to view their own legislators as agents of local interests. Specifically, they place high value on service to the district, communication with constituents, and, using Fenno's phrase, "home style"—the way a member deals with the home folks.[12] The historical high rate of return of incumbents—from both parties—suggests that citizens generally draw a rather sharp distinction between their own Representative and Senators and "the rest of those politicians." Some in Congress—and out of it—offer a pungent explanation for people making this distinction: it is argued that much of politics is about "taxing them and spending on us." From this it is argued that "my Congressman brings home the things we need, while all those other bums tax us to pay for 'pork' for those other people."

TWO WORLDS

Members must adapt to moving back and forth between the environment of Washington, where they are powerful and catered to, and the district environment, where they are seen as servants of local interests. Anecdotes related by members and a wealth of data underscore the enormous difference between the relationships members have with constituents compared with their relationships with lobbyists, staff, and executive branch officials. Constituents relate to lawmakers in a unique way that resonates with the core, central interests of the member. Some observers say that Congress is out of touch. In reality, the problem is not that members are insulated within the Beltway and out of touch with their constituents. Rather, one could argue that the problem is just the reverse; the necessity to keep the electoral connection strong forces a certain hypersensitivity on the part of legislators to the wishes of the home folks. As former Speaker Tip O'Neill said, "All politics is local," and, in the aggregate, this may well collide with reaching national interest decisions—such as reducing the budget deficit.

IMPLICATIONS FOR STAFFING

Finally, this duality of Congress is reflected in the way members staff their offices. Increasingly, representatives and senators have established a number of offices in their districts and states and have located sizeable fractions of their staff in these offices. Most constituent service functions are handled at the state or district offices. Much of what members do in Washington is also district oriented. In many offices the staff spends a good part of their time dealing with the policy concerns of local groups and correspondents.[13]

THE NEW CONGRESS: DYNAMIC AND FLUID

REFORMS AND POWER SHIFTS

In seeking to understand Congress, it is essential to grasp the fact that it is a fluid, dynamic institution. You cannot study it once and go away thinking you know what you need to know to understand its workings. Lawrence C. Dodd and Bruce I. Oppenheimer, in their continuing editions of *Congress Reconsidered*, have tracked the unfolding evolution of this complex institution.

As a result of reforms undertaken in the House in the early 1970s and the Senate in the mid-1970s, Congress was becoming a fragmented, decentralized institution, dominated by subcommittee government. The 1980s and 1990s have seen a quite different pattern. There has been little discretionary money for new programs and thus little opportunity for policy innovation by the authorizing subcommittees in Congress. Indeed, renewed emphasis on reducing the deficit and balancing the budget has caused budgets in many areas to constrict. Dodd and Oppenheimer suggest that "power has shifted toward the party leadership and a few elite committees that deal with essential annual legislation."[14] Republican House Speaker Newt Gingrich expanded the reach of the House leadership in the 104th Congress by tightly controlling the appointment of committee chairs and keeping a close eye on their agendas.

The Senate differs significantly from the House in that Senators tend toward greater autonomy and individualism. Because of its smaller size and collegial style, the Senate can better tolerate personal idiosyncrasies and operate in an orderly manner without strong leaders. Also, because rules are much less controlling in the Senate than in the House, the Senate can bring about change in more diverse, less visible ways than the House. As a body, the Senate serves as a check upon the "majority rules" flavor of the House. Its rules and customs virtually require compromises and accommodation to minority views and interests. Smaller size and longer tenure not only allow but encourage more informality than in the House. Senators get to know each other personally to a much greater extent than do Representatives. However, despite the six-year term, Senators now spend far more time campaigning than in earlier decades, and junior Senators have been much less inclined to take minor roles and to wait patiently for seniority.

In the House, the coexistence of subcommittee government and strong central leaders has created the potential for a dynamic fluidity in power arrangements. We seem to be in a period of divided power within Congress, where large swings in power can take place in response to external forces. Such swings, conclude Dodd and Oppenheimer, probably are most evident in the House. The Senate is better insulated from outside forces, but it nevertheless follows the House pattern to some extent.[15]

THE IMPACT OF DIVIDED GOVERNMENT ON CONGRESS

Democrats controlled the House of Representatives for forty years and the Senate for much of that period until the Republicans regained control of both the House and Senate with the 1994 elections. With Republicans controlling Congress and Democrats controlling the White House, a period of policy stalemate in a number of areas, reminiscent of the days when Democrats controlled Congress and Republicans controlled the executive branch, has continued. This state of affairs could, of course, change again at the choice of the voters.

Each passing year in the 1980s and until 1992 had seen the nation's budget deficit continue to mount. While the 1988 and 1990 elections produced no political outcome that could lead to addressing the budget problem in fundamental ways, the 1994 elections did. After a budget showdown in November 1995, President Clinton and Republican leaders agreed on some elements of a plan to balance the budget over seven years. Getting such an agreement in place might free legislators to concentrate on other areas. The ability of Congress to innovate and serve as an incubator for legislative proposals and policy initiatives that involve spending money has been greatly constrained. This ties in with the trend toward centralization and toward a focus on those committees that deal with budgets and appropriations. This has been styled the "fiscalization" of national policymaking.[16]

CHANGES IN MEMBERS

At least some of the changes just described are due to a large infusion since the 1970s of younger legislators in both the House and Senate, legislators who are dedicated to reform and who take a more activist approach to being a member of Congress. This is no longer an institution dominated, as it has been in the past, by aging elders and long-time office holders. There are many new faces: 52 percent of all members of the House of Representatives in the 104th Congress have served three terms or less and 55 percent of all senators are in only their first or second term.[17]

Both the Senate and the House are more egalitarian institutions than they used to be. Relatively junior members in both bodies now play important roles, often in an entrepreneurial mode comparable to that of their private sector counterparts. Such trends have opened up the political system to more participants and attracted more talented individuals, and less public business is conducted behind closed doors. In short, Congress has become more democratic, more accountable, and more open.

Yet these trends also have their negative overtones and serious costs. Former Congressman and New York University President Emeritus John Brademas, for example, has observed that "Congress is more and more becoming a place of independent contractors, each Member intent on constructing his record in a manner most pleasing to the eyes of his constituents or special interests but without regard to his responsibility to serve the national well-being." Brademas worries that if members of Congress pursue their individual goals and are unwilling or unable to follow leadership in support of common positions, "we will never be able to construct coherent policies to deal with genuine national problems." Another practical cost noted by Brademas and others is the greatly increased "time and effort required to get things done, not a trivial consideration in a period when the sheer volume of problems that government must address threatens to grow beyond manageability."[18]

At the insistence of reformers in both houses, power, in the form of subcommittee chairmanships (about ninety in the House and seventy in the Senate), has been distributed more widely.[19] Among the consequences of this power redistribution—and other reforms noted earlier—has been a more dispersed, more fragmented, and more open institution. Such is the case not withstanding the 1994 "Republican revolution" and a more centralized operation—especially in the House. One implication of all these changes is that it is easier than ever for outside interest groups and advocates to introduce their ideas into the legislative process and have them acted upon.

CONGRESS IN POLICYMAKING

Analysts of Congress and the presidency have devoted much time and attention to the roles and relative influence of the two branches in their exercise of political power. Walter J. Oleszek has said, "Much has been written about the growth of executive power in the twentieth century and the diminished role of Congress, but in fact there has been a dynamic, not static, pattern of activity between the legislative and executive branches. First one and then the other may be perceived as the predominant branch." But Oleszek and others have argued persuasively that at any given time of such appraisals, the descriptions often underestimate the other branch's strategic importance.[20] The American political system is best considered largely as a combination of congressional and presidential government. Yet Congress is elected separately from the President and has constitutionally based powers and a substantial degree of independent policymaking authority. Historically, this has been particularly true in a number of areas related to science and technology (e.g., environment, energy, health, R&D organization, science and mathematics education, and economic competitiveness).

Over the past two centuries, the House and Senate have used their constitutional mandate to devise their own sets of rules, procedures, and conventions, both formal and informal. As Oleszek notes, "these rules and conventions establish the procedural context for both collective and individual policymaking actions and behavior." More specifically, rules and procedures serve many purposes in Congress: to provide stability, legitimize decisions, divide responsibilities, reduce conflict, and distribute power.[21] To work effectively with Congress, it is important to be aware of the importance of rules and procedures and to acquire at least a passing acquaintance with those that may be relevant to your concerns. A splendid source for gaining such familiarity is Oleszek's *Congressional Procedures and the Policy Process.*

Under changes in the rules and procedures of recent years, Congress has dispersed power to many members and has opened itself to greater public observation and inspection. In today's Congress, many members, junior and senior alike, can exercise initiative and creativity in both lawmaking and legislative oversight activities. Overall, this means that the representative or lawmaking role of members has been enhanced under decentralization: members can easily obtain diverse and competing views, independent of the communication channels managed by the leadership. But decentralization and openness carry the usual price of more democracy: it frequently takes Congress longer to formulate public policies, compromises with the President are more difficult to achieve, and policy stalemate can result.[22]

In summary, Congress, in its policy-making role, displays the features, traits, and biases of its members, rules and procedures, structure, and constitutional underpinnings. It is bicameral with divergent electoral and procedural traditions. Congress is representative, with respect to both issue constituencies and geographical interests—especially the latter. It is decentralized, power is dispersed, and there are few mechanisms for integrating and coordinating its policy decisions—especially as they cut across traditional issue and organizational boundaries, within and outside Congress. Finally, it is often reactive, reflecting conventional or elite perceptions of issues and problems.[23]

One negative aspect of all of this is that congressional policymaking can often be overly driven by localism; afflicted with partial, piecemeal solutions when the problem calls for an overarching, national solution; saddled with ceremonial, "make people feel good" solutions that solve nothing; and frozen in a state of near paralysis pending the arrival of a crisis. However, history suggests that the upside, when the crisis finally arrives, is that Congress can respond quickly and, most of the time, appropriately. Moreover, Congress provides the place and the circumstances to hammer out consensus on controversial issues—taking months, years or decades, if necessary.

While consensus and compromise are being sought, Congress serves as ombudsman for citizens in dealing with their national government. It offers the most open access of any part of the national government: there is an opportunity for individuals to take their concerns and their problems to the very top of our political system. While this "hammering out" process is taking place, Congress may appear to be in disarray and paralyzed. The public and the media may be clamoring for action that Congress cannot produce. But the fact is that Congress is often doing a good job of representing and reflecting the fact that no consensus exists on an issue among the American people.

MAKING THE SYSTEM WORK

While the Constitution provides a marvelously constructed political system and a set of time-tested basic principles, it is itself flexible, adaptable, stretchable, and fuzzy in key places. These features also characterize the political system it established. As a result, the power relationships among the three branches of government are in a constant state of tension and dynamic flux. This requires:

- A tolerance for differences

- A central tendency to compromise

- Acknowledgment that times and conditions change

- A broad correspondence with the popular will

- A deep understanding of limits (e.g., how far to push one's views)

One troubling feature of the mid-1990s political scene is an apparently diminished appreciation for such guidelines by many politicians. Retiring members of Congress from both parties have remarked on a decline in civility, a lack of willingness to compromise, and little respect for other's views. Zealotry and single-mindedness in the pursuit of one's own approach can only make the situation worse, according to some who are choosing to leave Congress.

Over time, the balance of power within and among the three branches has changed—sometimes dramatically—due to national emergencies, correction of abuses of power, the changing role of the nation on the world scene, and the vast growth of our economic and technological enterprises. Above and beyond the formal processes of government, informal processes and communications are necessary to make the system work. In seeking to make the system work, eventually there is a unifying theme to the purposes of Congress and the presidency: ascertaining and implementing the people's will. Even the Supreme Court is not immune to this broad purpose. As Walter J. Oleszek says, "The Constitution, in short, creates a system not of separate institutions performing separate functions but of separate institutions sharing functions. The overlap of powers is fundamental to national decision making."[24]

ENDNOTES

1. Walter J. Oleszek, *Congressional Procedures and the Policy Process*, 3rd ed. (Washington, DC: CQ Press, 1989), p. 2. The discussion of basic constitutional principles closely follows Oleszek's treatment of them.

2. Roger H. Davidson and Walter J. Oleszek, *Congress and Its Members*, 3rd ed., (Washington, DC: CQ Press, 1990), p. 1.

3. Giovanni Sartori, "Representational Systems," in *The International Encyclopedia of the Social Sciences*, Vol. 13 (1968), p. 468.

4. Davidson and Oleszek, pp. 4-5.

5. Congress collectively is composed of 100 senators, 435 representatives, and a total of five delegates from the District of Columbia, the Commonwealth of Puerto Rico, and several territories.

6. Davidson and Oleszek, p. 5.

7. Ibid., p. 10.

8. Thomas J. O'Donnell, "Controlling Legislative Time," in Joseph Cooper and G. Calvin Mackenzie, eds., *The House at Work* (Austin, TX: University of Texas Press, 1981), pp. 127-130.

9. Ibid., p. 6.

10. Richard F. Fenno, Jr., *Home Style: House Members in Their Districts* (Boston: Little, Brown and Co., 1978), p. 104.

11. Davidson and Oleszek, p. 7.

12. Ibid.

13. Lawrence C. Dodd and Bruce L. Oppenheimer, eds., *Congress Reconsidered*, 4th ed. (Washington, DC: CQ Press, 1989), p. 421, 436.

14. Ibid., p. xi.

15. Ibid., p. 448.

16. Allen Schick in Dodd and Oppenheimer, p. 441.

17. Norman J. Ornstein, Thomas E. Mann, and Michael J. Malbin, *Vital Statistics on Congress 1995–1996*, (Washington, DC: CQ Press, 1996).

18. John Brademas, *Washington, DC to Washington Square* (New York: Weidenfeld and Nicolson, 1986), pp. 160-161.

19. Norman Ornstein in Dodd and Oppenheimer, p. 20.

20. Oleszek, pp. 4-5.

21. Ibid., p. 5.

22. Davidson and Oleszek, p. 305.

23. Ibid., pp. 374-377.

24. Oleszek, p. 2.

3 CONGRESS AT WORK

T he work of Congress can be divided into two broad categories: lawmaking, which includes budgets, appropriations, authorizations, other legislation, and oversight; and representation, or serving constituents, as well as building and reinforcing political support.[1] Understanding these two worlds is crucial to the person who hopes to influence a member of Congress. When a member listens to your opinion on the substance of a scientific issue, he or she is listening in a different language—the language of political nuance and potential political value. In discussing these two worlds in which Congress works, this chapter examines the growth in congressional workload, the organization of Congress and its staffing, the committee system, and the sources from which Congress obtains information and advice.

THE MEMBERS OF CONGRESS

PERSPECTIVE

Members of Congress often lock horns over how to make laws that best reflect their constituents' views and needs. Gridlock sometimes results as politicians vie to control policy and process, rhetoric and results. In November and December 1995 and into January 1996, a large part of the government shut down in showdowns over approaches to balance the budget and Medicare/Medicaid priorities. Bills that would have allowed the government to spend money were held up because of these policy disagreements. With no money appropriated for employees' salaries and agency operations, many government agencies closed until a political compromise could be reached.

Yet political gridlock in Washington happens not necessarily because of obstinacy on the part of Congress and the President, but because Americans have chosen a style of divided government. This is due in part to having checks and balances on powers among the three branches of government. It is also due to our inability, as a people, to reach consensus on how to proceed in solving deeply rooted national and global problems.

Many different parts of the nation and many powerful political and issue groups have dug in their heels, refused to consider the path of accommodation, and not tolerated a view different from their own on issues from deficit reduction to reproductive rights. Congress is reflecting or mirroring, perhaps all too well, this lack of consensus within the electorate. If Congress appears to be in disarray, it is largely because the electorate itself is in disarray.

Understanding who the members are, the work they do, and the climate in which they work can give a citizen scientist or engineer insight into how best to shape his communication with a member of Congress.

BASIC STATISTICS

The elections of 1994 proved to be a time of great change for the U.S. Congress. Not only did the House fall under Republican control after four decades of Democratic leadership, but a near record number of members retired or left to seek other positions. And as the elections of 1996 approach, it appears that the retirement record, set in 1978 with fifty-nine retirements, may be broken.[2] More than 52 percent of all members of the House of Representatives have been elected since 1990. In the Senate, 54 percent of members have served twelve years (two terms) or less. In other words, about half of all the seats in Congress have turned over during the past decade. Such data do not support charges that Congress is the bastion of entrenched, aging politicians who are out of touch with the people. It must be acknowledged, however, that such aggregate data do not reveal the great variety in election patterns throughout the country. Some seats are safe for incumbents, election after election; other seats are in a continuing "swing" status; others are close but tend to tilt toward one party; and all are subject to periodic upheaval.

The influx of new faces in Congress since 1990 has resulted in a slight decline in the average age of members from 54 in 1992 to 52 in 1994. The number of women in the House remained the same from 1992 to 1994 at 48, but the Democrats lost five women while the Republicans gained five. The Senate increased its ranks of women to eight with the addition of two new Republicans. The number of African-Americans in Congress remained constant at 39 Representatives and one Senator. Again, the Democrats lost one African-American House member while the Republicans gained one.

The number of members with a business or banking background grew to 186 in 1994, a 20 percent increase, while the number of lawyers decreased almost 6 percent to 225. As listed in *Vital Statistics on Congress 1995–1996*, six members are engineers, 11 have backgrounds in medicine, and 75 are educators. Of all members, 114 list public service/politics as their previous occupation. At least two House members of the 104th Congress are Ph.D. level scientists, one a chemist and the other a physicist.

A brief profile of the Senate shows there are no fewer than fourteen former state governors, several former mayors, and at least five Rhodes Scholars. Many senators had highly successful careers in fields outside of politics before coming to the Senate. The great majority have educations beyond an undergraduate university degree, and many of their advanced degrees come from the nation's premier universities. Very few senators are professional politicians in the sense of having devoted their entire careers to serving in a political office.

However, about a third of the current senators have served in the House of Representatives. About a third of all senators are millionaires, and some are exceptionally wealthy. While some of the wealth is inherited, much of it is self-made.

Not surprisingly, the House does not appear to be as attractive to former governors of states as the Senate. In the House there is only one former governor and one former lieutenant governor.

GROWTH IN CONGRESSIONAL WORKLOAD

Anyone seeking to influence or communicate with Congress must recognize that while communication lines are open, many people are using them. Those who want to get a member's attention must expect a lot of competition. The workload of Congress has increased tremendously over the past three decades. Davidson and Oleszek note that this workload, which was "once limited in scope, small in volume, and simple in content... has grown to huge proportions." These changes are in direct response to the changed character of Congress.[3]

Until the 1950s, Congress was largely a part-time institution, and members were paid as if they were doing a part-time job. For the first part of the twentieth century, Congress stayed in session only nine months of each twenty-four; the members spent the rest of the time in their districts or tending to their private business. Recently, Congress has been in session nearly all year except for occasional district work periods. The average senator and representative works at least an eleven-hour day while Congress is in session and often even longer in his or her state or district. Congressional staff members have comparable workdays. It once was expected that members and staff should have outside jobs; this is now prohibited, and indeed, time constraints make it impossible.

A variety of indicators may be used to measure congressional workload and its growth. Some of these include time in session, committee meetings, and floor votes. By such measures, the congressional workload has nearly doubled over the past thirty years. Committee hearings have so proliferated that members have conflicting schedules and cannot attend all the hearings of the committees and subcommittees to which they belong.

There have been downturns since the 1980s in some indicators such as numbers of bills enacted into law (see Figure 3-1), but these are a reflection not of a decreased workload but rather of its changed nature. There has been a shift to more "mega-bills"—particularly in the budget area—and an increasing emphasis on oversight of the executive branch and on investigations that often lead to a report but not to legislation. Apart from numerical indicators, overall legislative business has grown in scope and complexity as well as in volume. In the 1980s and early 1990s Congress dealt with many issues that once were left to state or local government or that were not considered to be within the purview of government at all. However, with the Republican takeover in 1994, the pendulum began to swing in the other direction, and there was a real focus on returning power to the states. Perhaps even more important, the issues of the 1990s are far more complex than those of earlier decades; more and more, they involve complicated interplay among a variety of factors—economic, political, social, and technological.

Another important dimension of the changed nature and consequent growth of congressional workload has arisen from the increased focus on constituent service and the related growth of the federal government's role in our daily lives. This increased focus is reported in academic studies of Congress, by members and staff themselves, and in such indicators as the amount of mail received and responded to by Congress. Davidson and Oleszek observe that "not only are constituents more numerous than ever before; they are better educated and served by faster communication and transportation. Public opinion surveys show that voters expect legislators to 'bring home the bacon' in terms of federal services and to communicate frequently with the home folks."[4]

Figure 3-1
Public Bills in the Congressional Workload
80th–103rd Congresses, 1947–1994

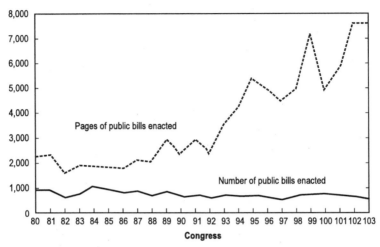

Source: Norman J. Ornstein, Thomas E. Mann, Michael J. Malbin, *Vital Statistics on Congress, 1995–1996.* (Washington, DC; Congressional Quarterly Inc., 1996).

ORGANIZATION OF CONGRESS

Formally, Congress does its work through individual member offices (in Washington and in the district or state), through committees, and on the floor of the House and Senate. The informal side of congressional organization goes well beyond this, however, and includes party committees, party caucuses, and a variety of other informal groups, which call themselves caucuses, coalitions, conferences, and so on.[5] These informal groups may be based on party affiliation (Republican Conference, Democratic Caucus), issues (New England Congressional Energy Caucus), geography (Northeast-Midwest Coalition), gender (Women's Caucus), or ethnicity (Black Caucus).

THE HOUSE LEADERSHIP
Although members of Congress consider themselves constitutionally coequal with the President, practical consid-

erations mean that some members are "more equal" than others. The role of Speaker of the House carries with it prestige and extensive power. The Speaker is recognized in the Constitution and designated as next in line behind the Vice President to succeed the President. With the selection of Republican Newt Gingrich as Speaker after the 1994 elections, the role of Speaker became even more pivotal. In the 104th Congress, Speaker Gingrich and his aides dictate the timing and management of bills—even in committee—often operating with sophisticated software to track, schedule, and manage the flow of legislation.

Although the Constitution does not require the Speaker to be a member of the House, all have been. Also, while formally elected by the entire House, as a practical matter the Speaker is chosen by the majority party. The Speakership combines policy and partisan leadership with procedural prerogatives. For example, the Speaker holds unique powers in scheduling floor business and in recognizing Members during floor sessions. Yet, in this modern Congress, the Speaker must be attuned to the Members and especially, but not exclusively, to the members of his own party. In the 104th Congress, Republican Members must also be closely attuned to the Speaker.

An elected majority leader is the Speaker's principal deputy. Both House and party rules are silent on the duties of the majority leader. In practice, the job is defined by the Speaker.

On the other side of the aisle, an elected minority leader is the titular head of the minority party. The functions of this position have included monitoring the progress of bills through Congress and forging coalitions with like-minded Members of the majority party. Another important role is the promotion of party unity and political leadership in seeking a return to majority status.

On the next rung of the party leadership ladders are the majority and minority whips.[6] Since the principal whips are elective posts, they have been seen as the path to majority or minority leader positions and the Speakership. The other whip positions (deputy, assistant, regional) are appointive. Whips in each party meet regularly to discuss strategy and issues. For both parties, whips aid the top leaders in gathering intelligence, encouraging attendance, counting votes, and persuading colleagues. Whips in the House often stand at the floor entrances to signal arriving colleagues on how to vote (thumbs up for yea; down for nay). Periodic whip notices are also sent out advising Members of upcoming floor agenda items and providing pertinent information.[7]

Subcommittee chairmen and ranking minority members of subcommittees are chosen by secret ballot within each committee by the respective party committee caucuses. Although seniority used to be the sole criterion for advancement to committee and subcommittee chairmanships, it is now only one of a set of factors used in filling these key positions. These days, other factors, such as loyalty to party policy positions, are also taken into account.

Significant organizations used by the leadership of both parties in carrying out their functions currently include the Democratic Steering and Policy Committees, the Republican Policy and Research Committees, and the House Rules Committee.

While the House of the 1990s is much more open and decentralized than it once was, it is still an institution where the leadership derives much of its power from holding the levers on the use of rules and procedures. Effective use of these rules and procedures permits a determined majority to achieve its policy or procedural objectives. But this can be taken only so far: ultimately, the leadership—both majority and minority—must persuade Members who represent different constituencies, values, and interests to support legislation before the House.

THE SENATE LEADERSHIP

The Senate is an institution suffused with individualism and independent operators, presenting dramatic challenges to the leaders, majority and minority alike, in performing their roles. Moreover, the rules and procedures of the Senate are much more flexible and much less structured than those of the House. For example, the Senate allows unlimited debate while the House places limits on the amount of time for debate. In addition, in the House, debate remarks and bill amendments must be germane to the issue or bill being discussed while in the Senate there are far fewer constraints concerning "nongermaneness" of what a senator has to say. Because of these differences, Senate leaders rely much more heavily on personal skills, persuasion, and negotiation than on rules and procedures to carry out the Senate's business.

The President of the Senate, as defined in the Constitution, is the Vice President of the United States. However, except for ceremonial or unusual occasions, he seldom presides, and he votes only to break a tie. The Constitution also provides for a president pro tempore to preside in the Vice President's absence. In practice, this role has been performed by the majority party senator with the longest continuous service, although other senators serve on a rotating basis. When a senator on the Senate Floor says, "Mr. President," the reference is to the president pro tempore. For practical purposes, however, the majority leader acts as the head of the majority party, as its leader on the floor, and as the leader of the Senate. In the same way, the minority leader heads the Senate's minority party and acts as its leader on the floor. Neither of these positions is mentioned in the Constitution; both are subject to elections with secret ballots at the beginning of each Congress.

As in the House, there are majority and minority whip systems. Their purposes include gathering votes and planning party strategy. Finally, as in the House, there are Senate party caucuses, committees, and informal groups. The undergirding organizational structures in the Senate are the Democratic Conference and the Republican Conference. Also important are the Democratic Policy and Steering Committees and the Republican Policy Committee.[8]

COMMITTEES IN CONGRESS

In the work of Congress, committees are at the center of things institutionally: in policymaking, budgets, revenues, investigations, oversight of federal agencies, and public education. While floor actions often refine the legislative products of committees, the committees are the means by which Congress sifts through thousands of bills and tens of thousands of nominations annually, along with considering issues and proposals by the hundreds.[9] It has been said that "on the floor is Congress for show, while in committee is Congress at work."

In 1995, the number of House full committees declined for the first time since 1955. The new Republican majority cut the number of House committees and subcommittees by 25 percent, eliminating three committees and implementing new rules restricting the number of subcommittees per committee to five.[10]

In a comprehensive study, Committees in Congress, Smith and Deering found that "committees still matter" despite widespread individualism and unstructured processes in the Senate and declining specialization of Members in the House, as well as diminished autonomy on the part of committees in both bodies. Members say that committees remain central to their personal goals, and they continue to judge carefully the value of particular committee assignments. For example, service on an appropriations committee is a powerful attraction, and appointments are vigorously contested. Also, most legislative activities of members revolve around their committee and subcommittee assignments.[11]

COMMITTEE ASSIGNMENTS

Committee assignments are central to the organization of Congress and to the ability of members to influence policy in areas in which they are interested. Such assignments are made under party rules and processes. Democrats and Republicans in the House and Senate set their own rules for assigning membership to committees, grading committees by level of importance. Membership on some of the committees, such as the Budget and Ways and Means Committees in the House, is exclusionary, meaning that if a Member serves on that committee they can serve on no others. Others, such as the Appropriations Committees in both the House and Senate, while not exclusionary, are considered more powerful and more desirable than others.

Senators usually divide their time and attention among a larger number of committee assignments than representatives. Moreover, senators have greater latitude than representatives to influence the agendas of committees other than those on which they serve. In 1995, the House took steps to limit the number of committee and subcommittee assignments per Member. The new rules limit Members to two standing committees and four subcommittees.

TYPES OF COMMITTEES

Committees vary considerably in importance and influence. The basic types are standing, select, joint, and conference. Even after a significant reduction in the number of committees in 1995, Congress still has a rather complicated organizational structure with 48 committees with more than 150 subcommittees, each vying for its place in the sun. This total does not include party committees and informal groups.

STANDING COMMITTEES

Standing committees are permanent congressional entities, established by law or by House or Senate rules. Such committees continue from one Congress to the next and process the vast majority of the daily business related to lawmaking, investigations, and oversight. From the thousands of bills introduced, committees choose a limited number to consider and send to the floor for possible enactment into law. Thus, they are also the burial grounds for most legislation. As of the 104th Congress, the House had 19 standing committees with 84 subcommittees. The Senate had 17 standing committees and 68 subcommittees.

The standing committees can be divided into five general categories dealing with budgets, appropriations, authorizations, revenue and taxes, and procedural/administrative matters. The Senate and House Budget Committees establish budget categories and overall targets for expenditures and revenues. In varying degrees, the authorizing committees enact laws providing legislative authority for the programs and agencies. The House and Senate Appropriations Committees (with thirteen subcommittees each) work out the details of appropriations for agency programs within these allocations and authorizations. The powerful Senate Finance and House Ways and Means Committees set tax and revenue policies and oversee most entitlement programs and all tax-related policy incentives. One other group of standing committees deals with internal congressional administration, operational rules and procedures, and administrative and judicial procedures and organization.

SELECT COMMITTEES

Select (or special) committees are supposed to last no longer than one Congress (a two-year period). However, some, like the House and Senate Select Committees on Intelligence, "just keep going" and take on the nature of standing committees.

JOINT COMMITTEES

Joint committees—of which there are four—have both senators and representatives as members. The chairmanship rotates between House and Senate. Joint committees are used when the House and Senate agree that the institutional interest of Congress as a whole should take precedence over the interests of either house. For the outside world the most important current joint committees are the Joint Economic Committee and the Joint Committee on Taxation; the other two are concerned with the Library of Congress and the Government Printing Office. It is worth noting that while Congress has a Joint Economic Committee, it has not chosen to establish a joint budget committee.

CONFERENCE COMMITTEES

Before a bill can be sent to the President for signature, it must be passed by each body in identical form. Conference committees are established temporarily to reconcile the differences between measures passed by the Senate and

House. Appointments to conference committees are of vital importance as this is where final agreements are reached about House and Senate differences, and the power a member can wield here is considerable.

AUTHORIZATION AND APPROPRIATION COMMITTEES

For those who work with Congress, one of the most important distinctions to bear in mind is the fundamental difference between authorizing and appropriating committees. In principle, the authorizing committees, of which there are many, produce bills that set policy, establish federal agencies and programs, and recommend budgets at certain levels. The House and Senate Appropriations Committees produce the legislation that actually funds the programs. Users of this guide will probably have more interactions with authorizing committees than other types. And while comparable interactions with the appropriations subcommittees are not as likely to occur, there is still a strong need to follow their activities closely, especially in these lean years of tightening budgets.

It is not unheard of for Congress to pass an authorization bill that is signed into law by the President but for which no appropriation is ever enacted. In this sense a program is authorized but does not really exist because the appropriations committees never allocated the money to implement it. Indeed, there are billions of dollars of such unfunded authorizations on the books. In part, this situation arises because there simply is not enough money to fund all enacted authorizations. It also arises because appropriations committees sometimes disagree with authorizing committees. Sometimes, the appropriations committees will fund a program regardless of whether or not an authorization bill has been passed.

GROWING TENSION

Over the years, tension has developed in Congress among the various types of committees over their respective roles and relationships. In many instances, new systems and processes were superimposed upon existing systems; the result has been a very complicated set of processes that are difficult to understand—even for many of those who work in Congress. Such a set of circumstances leads easily to power plays and competition among various members and committees.

STAFFING FOR CONGRESS

One of the most important aspects of Congress is staffing. As Fox and Hammond note, "Congressional staffs, both committee and personal, are an increasingly vital resource to Members of Congress." As Congress has changed, staffs have expanded and changed—in distribution and qualifications.[12] The substantial growth of congressional staff over the years is shown in Figure 3-2. (Although the current figures were not available for this chart, when the Republicans took over the House, they pledged to decrease staff by 25 percent.)

Generally, congressional staff members today are better educated, more professional, and possess a greater variety of skills and backgrounds than staff of earlier decades. There are far fewer "political hacks" who come into the system solely on the basis of their political connections regardless of competence. There is no typical staff person, but, in general, personal staff tend to be younger and somewhat more mobile than committee staff. While some people in key positions on the Hill have been there for years, many others view a congressional staff assignment as a way station in their career development and serve for only a few years.

Members are ambivalent about the role of staff—always depending on them but sometimes resenting this dependence. There is little doubt, however, that Fox and Hammond are correct when they say, "The key aspects of what makes Congress run—activity, communication, organization, and community—in large measure involve staff. Many congressional outputs can be traced back to staff activity, where they conceptualize, write, type, and finally communicate a message."[13]

PERSONAL AND COMMITTEE STAFF

It may be useful to distinguish between personal and committee staff, but it is nearly impossible to generalize on

Figure 3-2
Staff of Members and of Committees in Congress, 1891–1993

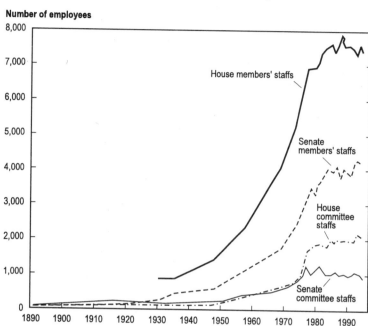

Source: Norman J. Ornstein, Thomas E. Mann, Michael J. Malbin, *Vital Statistics on Congress, 1995–1996.* (Washington, DC; Congressional Quarterly Inc., 1996).

such differences because of the wide range of organizational models and operating styles of the many hundreds of personal and committee offices. Also significant are the many differences between Senate and House styles and practices. With some oversimplification, here is a brief summary of distinctions between personal and committee staff:

■ In both the Senate and the House personal staff tend to be younger, less experienced, and more mobile than committee staff members. Also, personal staff tend to be somewhat more parochial in their outlook than committee staff. There are distinct and natural tendencies to assess the value or importance of an initiative in terms of its impact on the district or state as well as on the member's reelection prospects.

■ Personal staff must always keep focused on their boss's priorities, interests, and constituencies—even while working on committee business. Committee staff, while serving all committee members, must be especially attentive to the committee agenda and the chair or the ranking minority member.

■ The roles of legislative aides on personal staffs vary tremendously in their functions. Some merely monitor hearings, issues, and legislative initiatives. Others, who are perceived as speaking for their bosses, can work actively and cooperatively with committee and other staff associated with their boss's committees and their boss's issues. Alliances—temporary and permanent—are used frequently to achieve desired objectives.

■ Even as legislative aides, personal staff are necessarily more involved than committee staff with the constituent service functions of Congress.

■ Personal staff may well draft legislative amendments, committee, and floor remarks for their boss and make suggestions to committee staff on legislative strategies. However, committee agendas are generally prepared by committee staff—in consultation with the chair and sometimes with the ranking minority member.

■ Senate personal staff generally have wide latitude to play active roles in legislative initiatives. There are more barriers in the House to such activities.

■ In both houses, selected committee staff go to the floor. In the Senate, personal staff may occasionally accompany a Senator to the floor; this is not permitted in the House. And committee staff are mostly responsible for scheduling, planning, and organizing hearings, as well as for writing reports and drafting legislation.

The staffing and structure of members' offices vary considerably, not only in overall size and types of staff, but also in the way things work. There is usually one senior staff person, more or less in charge of the staff, often called the administrative assistant (AA) or chief of staff. (The title administrative assistant is a carryover from the days when each member's staff consisted of only a secretary or two.) A member's AA is usually the most powerful personal staff person. Some members work with many staff directly, while others prefer a more hierarchical approach through the AA.

Another common feature is the division of labor between case workers, who handle constituent services, and legislative assistants, who work on policy issues. There is often a legislative director (LD) who coordinates the work of the legislative assistants. The LD also helps to develop broad legislative strategies for members and performs committee-related work for those members who do not have committee staff assigned to them. Other important staff include a press secretary and a personal assistant to the member. The personal assistant handles scheduling and is often referred to as the appointments secretary or scheduler.

LEGISLATIVE AIDES

Legislative aides—personal and committee—can influence congressional decision making significantly. For example, staff have almost complete control over communications into and within a committee and personal office. Staff often have lead roles in identifying issues and developing legislative positions. Among their tasks are conducting research, gathering background data on specific legislative matters, and drafting legislation. They research and draft testimony, speeches, floor remarks, letters to constituents, and reports. In cooperation with other staff, they increasingly coordinate legislative strategy. They track a multitude of issues and events and provide succinct briefings to their member, often in the five-minute walk from the member's office to the floor. Finally, and perhaps most important, they offer their opinions and serve as a sounding board for their member or chairperson.[14]

As Figure 3-2 and Table 3-1 show, in 1993 there were just under 1,000 committee staff in the Senate and slightly more than 2,100 in the House. This makes a total of about 3,100 staff who are directly concerned with the business of government, that is, with congressional functions related to lawmaking, investigations, and oversight. While some may think this a large number, it is less than 10 percent of the number (about 40,000) often used by critics of Congress to describe a bloated staff system. Table 3-1 also notes that House personal staff number about 7,400 or just about 17 per office, since each member receives the same personnel allowance. In the Senate, personal office staffing amounts to about 4,100 and is allocated on the basis of state population: senators from states with small populations, such as South Dakota, get a far smaller allocation than do senators from large states, such as California. The overall growth in staff has leveled out since the early 1980s. The earlier growth was largely in response to growth in the roles and size of the federal government and to the increasing number of executive branch agencies. In one sense, staff equals power for Congress to legislate, investigate, and oversee the operations of government.

UNELECTED LAWMAKERS?

Members of Congress simply cannot handle the heavy workload on their own; they must rely extensively on what some have called unelected lawmakers. As Davidson and Oleszek have observed, "their influence can be direct or indirect, substantive or procedural, visible or invisible."[15] The discretionary agenda of Congress and its committees, and even that of personal offices, is greatly influenced and shaped by the staff. For example, senior committee staff members and staff directors have described in considerable detail how committee agendas are planned: the chair and ranking minority member will have their interests taken care of in the process, but so will the staff.

Table 3-1

Distribution of Congressional and Support Agency Staff

Congressional Staff[a]

	House	Senate
Member personal staff	7,400	4,138
Committee staff	2,147	994
Leadership Staff	137	100
Officers of the House/		
Senate Staff	1,194	1,165
Subtotals	10,878	6,397
TOTAL (House & Senate)	**17,275**	

Support Agencies[b]

Architect of the Capitol	2,500
General Accounting Office[c]	3,500
Government Printing Office	4,000
Library of Congress	4,800
[Congressional Research Service 760]	
Congressional Budget Office 226	
TOTAL (Support Groups)	15,026
GRAND TOTAL	**32,301**

a. Congressional staff figures are from 1993 and are based on data from *Vital Statistics on Congress, 1995–1996.*

b. Figures are from 1995 and are based on data acquired from each of the support agencies.

c. GAO estimates that 80 percent of its total workforce of 4,350 is connected with service for Congress. By the end of fiscal year 1996, the total staff level at GAO is scheduled to decrease to 3,500.

Throughout the many stages of congressional policy-making, staff members play an active part. Policy proposals arise from many sources inside and outside the government, and congressional staff members are positioned so as to be able to advance or hinder these proposals. Staff actively engage in negotiating with members, lobbyists, outside interest groups, and executive branch officials on issues, legislative language, report language, and political strategy.[16] Staff members are deeply involved in preparing committee agendas (legislative and oversight), planning hearings, recruiting witnesses, drafting reports and legislation, participating in drafting amendments for committee mark-up sessions, and even (in limited numbers) accompanying members to the floor when their legislation is being considered.

In contrast to many other organizational settings, the staff organization pyramid in congressional offices is often flat, and many of the professional staff have direct access to the member. Organizational norms such as promotion on the basis of merit, seeking formal professional recognition, following standard career patterns, and rigorous selection procedures are not characteristic of the congressional setting. Rather, personal loyalty, persistence, deference, special rules of courtesy, maintaining a low public profile, and a lack of concern with formal assigned duties and tasks seem to be associated with both committee and personal staffs. There is very much the sense of a personal team, whether it be a personal or a committee office.

PROFESSIONALISM AND PARTISANSHIP

Many staff are necessarily partisan both in their orientation and in their congressional activities. Since most staff are appointed by a partisan, they may be expected to and, most often do, reflect the partisan orientation of their patron. Professional experts coming into the congressional staff system or working with it must be aware of and sensitive to these features of Congress.[17]

HEARINGS

The budget committees, appropriations committees, and authorizing committees—as well as joint and special and select committees or their subcommittees—all hold hearings related to legislation being considered to conduct oversight of executive branch agencies and programs. The character of the hearings, and the range of testimony that is heard, varies widely.

BUDGETS AND APPROPRIATIONS

Money-related congressional activities are very much an inside game. Opportunities are limited for the public to participate in the work of budget committees and appropriations committees and their associated hearings. Agency representatives appear to explain and defend their programs and the President's budget request, however, outside or public witnesses presenting expert testimony are rare.

To the extent that public witnesses are involved, it is most often through "public days" set aside by the committees. The budget committees do extend invitations to some witnesses, but the prevailing practice in the appropriations areas is to require individuals and organizations to request time, typically for only a few minutes, and generally as a supplicant rather than as an advisor. In short, there is only a very limited requirement in these types of congressional hearings for expert information and advice. When it is desired, the committees tend to acquire information informally and through special reports or investigations conducted by investigators on loan from various agencies.

LEGISLATIVE PROPOSALS

There is a substantial requirement for public testimony involving expert advice and information in congressional hearings before the authorizing committees. That has created a large and growing demand for information, analyses, and expert advice on virtually every important national and international issue—as well as for many of limited import or consequence. Divided government, with different parties controlling the Congress and the executive branch, also leads to higher levels of policy conflict than is the case when a single party controls both Congress and the presidency. Another factor is that there are many more players and a decentralization of power among those players than in the 1960s. One of the results of the congressional reforms of the mid-1970s was the distribution and decentralization of power among subcommittee chairmen and substantially increased staffs.

OVERSIGHT

At least equal to the legislative demand is the very large requirement for expert advice, analyses, and information in the form of public testimony for oversight activities performed by Congress. Oversight can be divided into four categories:

- Committees involved with authorizing legislation are required to review the programs and operations of federal agencies within their jurisdiction and to recommend appropriate corrective action for problems.

- The appropriation subcommittees in the House and Senate have financial oversight responsibilities. For revenue-related activities, the Senate Finance Committee and the House Ways and Means Committee undertake oversight activities.

- A broad range of investigative responsibilities are assigned explicitly to the Senate Governmental Affairs and House Government Reform and Oversight Committees. Their mandates are not confined to any particular agency or set of issues.[18]

- The authorizing committees perform wide-ranging investigations under a mandate from the Legislative Reorganization Act of 1946, which calls for Congress to perform "continuous watchfulness" over the agencies under the jurisdiction of various committees.

In recent times, congressional committees have used their investigations and oversight powers increasingly in a large number of areas. Then-Senator Harry Truman's investigation of procurement fraud during World War II was a major stepping stone to the Vice Presidency and ultimately, the Presidency. Richard Nixon used the Alger Hiss case in much the same way. Other members of Congress, over the years, have combined oversight investigations with an active media interest in selected areas, whether it be alleged scientific fraud or charges of misuse of indirect costs at universities, perjury at the Environmental Protection Agency, or misinterpretation of global environmental data. Such oversight activities may well advance individual members' agendas, yet other members see these activities as a fully legitimate use of congressional power. Recent oversight hearings and investigations into scientific misconduct and university management of research funds are seen in this light.

INFORMATION AND ADVICE FOR CONGRESS

Information and advice are available to members of Congress from a wide range of sources. Some are internal, such as other members, staff, and congressional support agencies such as the General Accounting Office or the Congressional Research Service. Others are external, including the executive branch, associations, interest groups and lobbyists, private individuals, and the media. Recent studies suggest that members are relying more heavily on executive branch personnel, the congressional support agencies, and interest groups and less on consultants and volunteer experts.[19]

INTERNAL SOURCES: OTHER MEMBERS AND STAFF

There is a vast, uncharted flow of information and advice among the members themselves. Over time, members develop their own personal networks, along with assessments of who is reliable, trustworthy, and knowledgeable and who is not. They depend on these networks and contacts for everything from advice on election campaigns to what to do about a given issue. Sometimes their staff aides know about these contacts, but often members keep both their sources and the specific advice private. The House and Senate floors, the cloakrooms, and even the gymnasiums (from which staff members are excluded) provide settings for such conversations.

An important implication for outsiders seeking to work with Congress is that some members have more status than others and are more influential in certain policy areas then are other members. Members cannot have expertise in all areas, so they turn to trustworthy or influential colleagues as necessary. This status is independent of formal organizational roles and depends more on how a member is viewed on a personal level by his or her colleagues. Assessing how influential a member is in these terms is not easy but should be part of your intelligence-gathering operation. Putting it bluntly, you reduce your chances of success if your main contact is not well thought of or is seen as having little influence within Congress. Identifying your issue too closely with a member held in low regard by colleagues may be the kiss of death.

Although members are sometimes ambivalent about their dependence on staff, there is no question that congressional staff constitute one of the main sources of information and advice for members. Levels of influence vary widely among staff members, however, and sensitivity is required in choosing which ones to work with.

CONGRESSIONAL SUPPORT AGENCIES

The three congressional support agencies, Congressional Research Service, General Accounting Office, and the Congressional Budget Office, have become highly integrated into the operations of Congress.[20] Their nonpartisanship, objectivity, and responsiveness to the requests of members make them valuable resources that members hold in high esteem—although each agency has encountered tensions and even hostility from time to time. One explanation of members' overall positive appraisal for the agencies may lie in an observation by Davidson and Oleszek:

> Unlike committee or personal aides, these agencies operate under strict rules of nonpartisanship and objectivity. Staffed with experts, they provide Congress with analytical talent matching that in executive agencies, universities, or specialized groups.[21]

These agencies are not chartered to provide information or analysis to the executive branch or to the public. Even so, many of their products are widely available formally and informally.

Congressional Research Service

The oldest and most widely used of the congressional support agencies is the Congressional Research Service (CRS), a unit of the Library of Congress. It provides an extensive array of research services, preparing analyses, monitoring issue areas and gathering pertinent data, and preparing legislative materials. Established in 1914 as the Legislative Reference Service, it focused initially on the preparation of law indices and digests and continues that service as well. CRS is one of the two support agencies—along with the General Accounting Office—that can provide direct assistance to both members and committees, not just to committees. Both of these agencies occasionally assign individuals on detail (or loan) to congressional committees for special purposes such as preparing a major report, planning a major hearing, or conducting an investigation.

The Legislative Reorganization Act of 1970, which changed the agency's name to the Congressional Research Service, also gave it a new emphasis on policy research and analysis. Since 1970, CRS's staff has grown from 200 to nearly 900, and a number of new units have been created, including a Science Policy Research Division. A sizeable number of CRS staff members come from academic backgrounds. More and more, they operate in a "think tank" mode for Congress.

CRS offers variety and versatility in its services. In one sense it reflects its library origins by providing quick responses to thousands of congressional requests annually for factual information. In another way, its policy research and analysis roles are illustrated by the publication of a large number of reports. These include the widely-used *Issue Briefs*, reports on many topics important to Congress, a *Digest of Public General Bills and Resolutions*, and summaries of major legislation considered by Congress. Technically, CRS reports are not available to the public, but, in practice, many of them are distributed outside Congress through members' offices.

The short preparation times, urgent deadlines, and confidentiality of many CRS products often preclude a formal outside review. To compensate, CRS has established a rigorous internal review process. First, a report is reviewed by peers within the author's division and, if appropriate, by analysts in other divisions for accuracy and analytical quality. After this peer review, the report is given a division-level review to ensure that it meets division standards of technical accuracy and congressional needs. Next, the Review Section of the

Office of Policy reviews the document for compliance with CRS standards of balance and objectivity, as well as for responsiveness to congressional needs, clarity, and timeliness. If the Review Office believes that additional peer review is necessary for technical content, it has the authority to request such a review.

When time and confidentiality considerations permit—for example, for reports and issue briefs that are written in anticipation of congressional requests—outside review is strongly encouraged. Usually such review is initiated by the analyst, but division management and the Review Office also have the authority to request such a review.

In addition to supplying written products, CRS briefs members and their staffs, analyzes issues, and holds numerous seminars for members and staff on a wide range of subjects. Finally, CRS works with committees in helping to establish their agendas, in suggesting areas they might want to investigate, and in providing checklists of legislation under the jurisdiction of various committees that requires revisiting or reauthorization.[22]

General Accounting Office

The General Accounting Office (GAO) is Congress's premier field investigator, domestically and internationally.[23] GAO was established in 1921, together with the Bureau of the Budget (now the Office of Management and Budget). Directed by the Comptroller General, who is nominated by the President and confirmed by the Senate for a fifteen-year term, GAO, as described by Oleszek, "conducts audits of executive agencies and programs at the request of committees and members of Congress to make sure that public funds are properly spent."

The agency focuses mainly on eliminating waste and fraud in government programs and improving program performance.[24] GAO maintains a staff complement of about 4,350 and issues more than 1,000 reports a year. The Comptroller General may assign teams of investigators to work on investigations extending over a number of months or years in close coordination with congressional committees.

GAO reviews can lead to congressional hearings, enactment of legislation, and significant administrative changes in the ways that executive branch agencies do business. Under the Legislative Reorganization Act of 1970, GAO was given an expanded role in providing oversight assistance to Congress. The Congressional Budget and Impoundment Control Act of 1974 propelled the GAO even more deeply into program evaluation, obtaining program data, and assisting in congressional oversight.

The products of GAO's reviews can take several different forms, including testimony by GAO representatives before congressional committees; oral briefings for members of Congress and staff, particularly on the progress of requested review; and written reports addressed to Congress, a requester, or an agency. Written reports range from strict statements of the facts to detailed analyses of the data obtained, including conclusions and recommendations. They are subject to stringent quality review procedures within the agency. Federal agencies and other parties affected by or related to GAO reviews are often given the opportunity to comment on draft reports—especially when the issues are sensitive or controversial or include significant recommendations for action by the agency head or Congress.

An especially important aspect of all GAO products is the independent stance of the agency. The Comptroller General is appointed for a fifteen-year term for just this reason—independence. In seeking to provide useful and credible analyses and information to Congress, the Comptroller General and the agency insist on planning, performing, and reporting their work independently and objectively. Thus the GAO maintains discretion in determining how and by whom the audit or evaluation work is to be performed. This discretion extends to deciding what is to be included in all of the GAO's products—oral and written.

Congress has directed that GAO reports be given wide distribution, in contrast to the limited distribution of CRS products. GAO's workload has grown enormously during the past two decades. A growing focus on "the budget as policy" has led to more interest and willingness on the part of members and staff to rely more on the sophisticated program evaluation and data-gathering techniques of GAO.[25] However, while roughly four-fifths of GAO's workload is directly for Congress, the rest is related to other purposes, including the GAO serving as a major audit arm of the federal government.

The special tenure and high status of the Comptroller General provides the agency with a significant degree of independence from both Congress and the executive branch. GAO reports often put forth controversial policy recommendations. It is no accident, then, that the GAO and agencies of the executive branch often end up in disagreements and confrontations over findings, conclusions, and recommendations arising from GAO investigations.

Congressional Budget Office

As the budget has gained in political significance in recent years, Congressional Budget Office (CBO) reports and analyses have received increasing attention in Congress, in the media, and in government in general. CBO provides economic and budgetary information in support of the congressional budget and legislative processes. The subject matter of the agency's work is rather broad, however, given that the budget of the federal government covers a wide range of activities and plays a major role in the U.S. national economy as well as the international economic scene.

In contrast to GAO and CRS, the Congressional Budget Office does not respond to individual member and staff requests. Its lines of communication are with the House and Senate committees. With a staff of about 225, it publishes reports on the budget, including scorekeeping; estimates tax receipts and government expenditures; makes economic forecasts and projections; estimates the cost of proposed legislative proposals (not always to the liking of their sponsors); and conducts background studies and analyses policy.[26] (Scoring, or scorekeeping, refers to the analysis CBO undertakes to ensure that the cost or revenue figures incorporated in a proposed budget or program are accurate.)

Like the General Accounting Office, CBO is frequently asked to testify before congressional committees. While the testimony is most often in connection with an ongoing or completed study, sometimes special analyses are prepared for such committee appearances. However, the most important CBO product is very likely its annual report to the House and Senate Budget Committees. One component of this report is a document providing economic and budget projections for the next five years; the other component is a plan for reducing the budget deficit.

CBO was established under the Congressional Budget and Impoundment Control Act of 1974 and was designed to be an integral part of a new approach to dealing with budget activities in the congressional setting. At least in part, it was intended to be the congressional counterpart to the Office of Management and Budget (OMB) in the Executive Office of the President.

The Congressional Budget Act was intended to strengthen the ability of Congress to deal with the federal budget and to restore the balance of budgetary power, which a number of analysts and participants believed had tipped too much toward the executive branch.[27] As part of this new framework, CBO is intended to give independent budgetary assistance, economic analysis, and policy analysis to Congress. CBO has established detailed procedures related to preparation of its products and maintenance of quality standards. These include an extensive set of detailed guidelines and questions to be addressed, internal and external reviews, extensive coordination among CBO units, and full clearance by the director's office before a report is released publicly.[28]

CBO has become a force not only in congressional budgetary affairs but also on the national scene as well. Its analyses of and reports on important policy issues have become essential sources for those inside and outside Congress. In recent times, CBO has become a central player in budget deficit reduction discussions through its projections about different proposals. The overall tone of CBO operations more closely resembles the hard-nosed, skeptical fiscal conservatism of the OMB than it does the expansive, program-initiating orientation of some congressional authorizing committees.

Office of Technology Assessment

From its establishment in 1973 until its demise in 1995, the Office of Technology Assessment (OTA) and its advisory panels of experts studied complex policy questions with scientific and technical implications. OTA was abolished after Republicans took control of both the House and Senate after the 1994 elections, as part of a reduction in the size of congressional staff. Despite the fact that its work and its reports were widely respected outside and inside Congress, OTA lacked the political clout it needed to fend off efforts to shut it down. At the end of September 1995, the one congressional agency solely devoted to science and technology policy closed its doors.

EXTERNAL SOURCES

Executive Branch

Congressional committees often correspond, at least approximately, with the jurisdiction and functions of one or more Executive Branch agencies. Examples include the Agriculture, Defense, Interior, and Labor Committees in both House and Senate. This correspondence can, over time, result in close working relationships among congressional committees, a member's personal office (especially a chairman or subcommittee chairman), the agencies overseen, and the interest groups most directly affected by the agency.

Scholars in political science and public administration have long studied these types of relationships. The arrangements have been characterized as the "so-called iron triangles— a shorthand term that embraced a wide variety of relationships." And whatever form they took, they entailed a large number of policy arenas.[29] The lines of communication between congressional and agency staff and between members and top agency officials are essential sources of information for both sides—even if they are at times a cause of anger in the White House, whether in Democratic or Republican hands.

For the outsider, these lines of communication mean that it is not always necessary to contact Congress directly in pursuing your interests. It may well be preferable to work indirectly, through an agency staff member who already has the access, trust, and skill in maneuvering through the turbulent waters of Capitol Hill.

Interest Groups

Interest groups, including lobbyists, play an active part in congressional activities. Interest groups perform important functions: informing Congress and the public; stimulating public debates on key issues; and making available to Congress all sorts of information and points of view on proposed legislation and oversight activities.[30] Specifically, Congress can look to interest groups for draft legislation, analyses of competing proposals (especially those of their adversaries), draft speeches, answers to questions (usually provided promptly), questions to ask at hearings, and a host of other things. In short, an interest group that is "plugged in" can operate in some ways as an extension of a member's or committee's staff.

While it is true that interest groups can provide a variety of benefits to those in Congress, a key challenge for members and their staffs is to use the information and assistance provided by interest groups without becoming bound to their wishes and specialized agenda. In a direct warning, E. Douglas Arnold states that in regard to governmental resources, "interest groups usually have their own ideas about proper allocation, and they seldom coincide with Congressmen's predilections."[31]

Other troubling aspects of the participation of interest groups in our political system derive from the confluence of the increasing costs and the consequent growth in the role of money with the diminishing influence of political parties in election campaigns. Special-interest groups have increasingly stepped into the breach. As John Brademas has suggested, "Such groups often concentrate on single issues about which they feel strongly, and they attempt to focus a congressional election solely on those issues."[32] This preoccupation with single issues is surely one of the more worrisome features of the U.S. political scene.

For better or for worse, with more than 2,000 special-interest group offices in Washington, this feature of our political system is solidly in place. Of course, not all interest groups are equal in terms of access or influence. Oleszek's conclusion, shared by many members and staff, is that an interest group's ability to influence congressional activities is based on several factors:

- The quality of arguments and information

- The size, cohesion, intensity of the organization's membership and the ability to marshall them

- The group's ability to develop alliances—temporary or permanent—with other organizations

- Its financial and staff resources

- The vision and shrewdness of its leaders[33]

Political action committees (PACs) have been established by various interest groups to raise and contribute funds to political campaigns. The continuing escalation in the costs of running for Congress has led many members and candidates to turn to special-interest groups, large numbers of which are willing to supply campaign money. The 1980s and 1990s have seen a mind-boggling growth in the number of PACs and in the amount and significance of their donations. This development was not foreseen when limitations were placed on individual contributions as part of the 1974 post-Watergate political reforms. John Brademas states that "to observe physical evidence of PAC power, one need only walk by a congressional committee room when legislators are writing the final version of a bill to see the host of lobbyists watching with interest and reporting every move that Representatives and Senators make."[34] However, the PACs themselves are watched carefully by the media and various self-appointed interest groups since, as part of the above-noted reforms, their activities are part of the public record.

The Media

No single influence is more important to Congress than the media. To put it simply, members and staff live, in direct and important ways, by what they read, watch, and hear in the media—as well as by what they create for the media. This involvement ranges from current newspaper headlines and the nightly television news to intensively analyzed issues in small-circulation but influential journals.

There are a few periodicals, journals, and newsletters (see Appendix C) which cover congressional action on science and technology issues. In addition, many professional societies and associations (see Appendix F) have newsletters that provide information on legislative action on issues within their particular discipline.

ENDNOTES

1. An important role of the Senate, the "advice and consent" function for presidential appointments and international treaties, is not discussed in this guide.

2. Much of the statistical data cited in this section is drawn from Norman J. Ornstein, Thomas E. Mann, and Michael J. Malbin, *Vital Statistics on Congress, 1995–1996* (Washington, DC: Congressional Quarterly Inc., 1996).

3. Roger H. Davidson and Walter J. Oleszek, *Congress and Its Members*, 3rd ed. (Washington, DC: CQ Press, 1990), p. 3. The summary treatment of congressional organization in this chapter follows that of Davidson and Oleszek.

4. Ibid., pp. 31, 34.

5. The Constitution says nothing about political parties, but they were devised early in the history of the Republic as a practical matter to deal with political organization.

6. The term whip was adapted from usage that arose during the eighteenth Century in the British Parliament—as are many other procedures, customs and practices of the U.S. Congress. Originally used in fox-hunting, the whip kept the pack of hunters and dogs under control. The sense of the adaptation seems rather obvious, although a physical whip is no longer used—despite the occasional urge of a modern congressional whip to have such a capability.

7. Davidson and Oleszek, pp. 159-166.

8. Ibid., pp. 167-176.

9. Ibid., p. 195.

10. Ornstein, Mann, and Malbin, pp. 114-115.

11. Steven S. Smith and Christopher J. Deering, *Committees in Congress* (Washington, DC: CQ Press, 1984), p. 271.

12. Harrison W. Fox, Jr. and Susan Webb Hammond, *Congressional Staffs: The Invisible Force in American Lawmaking* (New York: The Free Press, 1977), p. 1.

13. Ibid.

14. Ibid., p. 2.

15. Davidson and Oleszek, pp. 218-219.

16. Ibid.

17. Fox and Hammond, pp. 157-159.

18. Walter J. Oleszek, *Congressional Procedures and the Policy Process* (Washington, DC: CQ Press, 1989), pp. 263-264.

19. Fox and Hammond, p. 122.

20. For a more in-depth discussion of these agencies, see Carnegie Commission on Science, Technology and Government, *Science, Technology and Congress: Analysis and Advice from the Congressional Support Agencies* (Carnegie Corporation of New York, New York, October 1991).

21. Davidson and Oleszek, p. 230

22. Fox and Hammond, pp. 131-132.

23. Oleszek, p. 271.

24. Ibid.

25. Fox and Hammond, p. 133.

26. Fox and Hammond, pp. 134-135.

27. *A Profile of the Congressional Budget Office* (September 1990), pp. 1-9.

28. "CBO Manuscript Procedures," in *CBO Administrative Manual* (Washington, DC: Congressional Budget Office, n.d.), pp. 3-10.

29. Roger H. Davidson in Thomas E. Mann and Norman J. Ornstein, *The New Congress* (Washington, DC: American Enterprise Institute, 1981), p. 105.

30. Ibid., pp. 136-137.

31. E. Douglas Arnold in Mann and Ornstein, p. 280.

32. John Brademas, *Washington, DC to Washington Square* (New York: Weidenfeld and Nicolson, 1986), p. 157.

33. Oleszek, p. 41.

34. Brademas, pp. 158-159.

LEARNING TO
4 WORK WITH
CONGRESS

This chapter offers guideposts on how to work more effectively with Congress in influencing its work as it may affect you or your organization. Remember, doing so is not a privilege; it is your right. But exercising your right effectively takes skill, knowledge, and practice. Formal and informal meetings with members and staff, telephone contacts, correspondence, and contacts with state and district offices are all discussed in this chapter. Hearings and testimony are covered in chapter five.

Many of these pointers are based on the discussions of the culture and workings of Congress in chapters two and three. In addition, these suggestions draw on the experiences and opinions of members of Congress, their staffs, and skilled professionals who work with Congress, much of which came out of questionnaires and personal interviews conducted expressly for this guide. In short, this advice is based on experience, some bitter and some sweet.

One important goal to keep in mind is to become a recipient of congressional requests for information or assistance. This is reflected in what one senior staff person said: "I live by my Rolodex." This individual and others emphasized how much they use the telephone in contrast to reading letters or reports. Members and staff operate within complex networks of information flows. Over time, each member and staff member develops a network of individuals on whom he or she comes to rely for advice, information, suggestions for prospective witnesses, evaluations of reports, assessment of people, and so forth. If you can become a valued source based on your performance, reliability, and credibility, then your opinion likely will be welcomed on scientific or technical issues.

17 CARDINAL RULES FOR WORKING WITH CONGRESS

Whatever mode you use for working with or contacting those in Congress, your overriding concern should be: How can I improve my chances for communicating my ideas successfully and getting them accepted? In operational terms, this means you should keep to the following guidelines:

1. Convey That You Understand Something about Congress.

A recurring complaint among members of Congress and their staffs is that so many who come to see them seem to know so little about Congress. Members and staff don't expect you to be an expert on Congress, but they do appreciate (and have more respect for) those who display an awareness and understanding of what is going on—particularly with regard to the conditions members and staff face. Among other things, these conditions include severe time constraints, competing demands for legislative and budget priorities, and the imperatives of reelection. Citing what may be an extreme case, a staff member explained why one visitor received a negative reception: "This guy didn't realize that representatives have to face an election every two years!"

2. Demonstrate Your Grasp of the Fundamentals of the Congressional Decision-making System.

Members and staff say that one of the most difficult things to get scientists and engineers to understand is the tough reality faced by members in balancing competing interests, building working alliances, and achieving acceptable compromises. Among their comments are that "scientific elites don't acknowledge other legitimate interests"; "there is a lack of understanding that they are in a competition like everyone else"; and "scientists are perceived as just another constituency." Finally, as one staffer pointed out, there is "a frequent misperception that a member will vote against one of his or her constituencies if only you will give them the correct facts." Unlike science, politics can't be reduced to empirical facts and figures. Indeed, it is rare that an initiative is not substantially modified through compromises and trade-offs before a final policy decision is made or a law is enacted. This means that you may lose even if you have a good case and a good relationship with the member. It also means that you should not take it personally and should keep trying. Persistence can pay off.

3. Don't Seek Support of Science as an Entitlement.

This may seem obvious, but it is a problem that occurs with sufficient frequency to require highlighting. Members and staff react negatively when they are presented with arguments in support of science that they see as being cast in "entitlement terms." In their words, scientists and engineers should not "convey an attitude of being inherently deserving in contrast to other seekers of the public largesse," and support for science should be "presented in terms of helping to meet national needs, or to achieve societal goals, not as an entitlement owed to scientists."

4. Don't Convey Negative Attitudes about Politics and Politicians.

Even if you have some inner, private views that are less than flattering about politics and politicians, keep them to yourself while working with Congress. It is the kiss of death to be perceived as having a "holier than thou" attitude, or as one staff member put it, to "convey that the purity of the scientific profession puts you 'above all of this.'"

5. Perform Good Intelligence Gathering in Advance.

Intelligence gathering involves learning at least the basics about the member, committee, or staff member you are contacting. As one staff member exclaimed, "Can you believe this person didn't even know which party my boss belongs to?" A good one-stop source is Congressional Quarterly's *Politics in America* (see Appendix C); other sources include hearing records, speeches, floor statements, and conversations with Washington friends who are knowledgeable about Congress. In addition, one staff member suggested that, "Too many people make a serious mistake by not leveraging or using the Washington offices of trade associations and companies."

Begin by learning where a member comes from (state/district), their committee assignments and professional background, where they stand politically on various issues, and how they fit into the congressional power structure. Try to learn if the member already has a view on your issue. As one senior staff person said, "Know what is on the member's mind in terms of recent concerns. Check recent hearings and floor debates." For staff, there is less published information, but it is still possible to get reasonably accurate profiles by making a few telephone calls to Washington friends, agency staff, association staff, and the office of your senator or representative, and by consulting the *Congressional Staff Directory*.

Staying in touch with developments is an important part of gathering intelligence. A good daily newspaper or weekly news magazine can keep you up to date on what Congress may be engrossed in at the moment. If a member is spending most of his or her time worrying about the budget or about foreign affairs, your recognition of that fact is important. Your sensitivity to such developments will smooth your road and perhaps your conversation. To stay on top of specific issues in Congress, you may want to read the *Congressional Quarterly Weekly Report* or the *National Journal* or check on developments through an on-line electronic subscription service such as Politics USA.

■— 6. Always Use a Systematic Checklist.

One good way to ensure careful, complete advance preparation is to write out a good checklist. Whether for participating in an elaborate meeting, presenting a statement in a hearing, or making a simple telephone call, prepare carefully. Think carefully about what you will want to leave at the meeting. This should include, according to a senior staff person, things right down to your business card with the date and a brief mention of your meeting topic. It was pointed out that "hundreds of cards get collected, at meetings, at receptions, and so on, and it is easy to forget six months from now how and why one has a particular calling card." Put together a simple, clear summary paper of one or two pages—including a brief background—that can be left after the meeting. Know what you want to say, and know when you've said it. Practice in advance with a dry run of your presentation. Demonstrate by what you say and how you present it that you are well organized and worth listening to. Such an approach will help you to plan better and to track your progress during a meeting or a telephone conversation. You are much less likely to get lost or to forget an important point.

7. Do Your Homework on the Issue or Problem.

It's obvious that you should know the technical side of your issue. Not as obvious, perhaps, is the importance of translating your message into terms relevant to Congress. Know which bills (if any) are pertinent. Know which committees are involved and what they are doing (e.g., holding hearings, planning hearings, and holding the issue in "deep freeze"). Know which other members are involved and what their views are. Tie your issue to member interests if possible. Look for connections between your issue and the member's interests. Such connections might be his or her legislative interests, they might be related to constituent concerns. One senior staff person said that, although it might not always be possible, you should try to "say why your proposal is important to the member's state or district, how current efforts are helpful that way, and why your proposal would be good for the member's constituents." You are on your way if you can clearly show the member how he or she can gain by going along with you. Finally, as one staff person said: "Try to have something to offer—good advice or useful contacts for additional information, for example." Use concrete examples as often as possible. Congressional people tend to be oriented to examples and anecdotes rather than abstractions or broad generalities. Play to this as much as possible without distorting your case. Staff members are always looking for "nuggets" to put into member speeches, floor remarks, and committee hearing remarks. Help them out if you can. As one staff member observed, "Concrete examples seldom hurt and most often they help."

8. Timing Is Vital.

All too often, the message may be great, but it is useless if the timing is all wrong. Keen judgment is required here. Weighing in too late with your opinion can mean the legislative train has left the station. As one committee staff director put it, "It was a good set of suggestions, but we'd already reported the bill out of committee two days ago. They thought we could fix it on the floor. Well, maybe—sometimes. But they should have come three months ago when it was still in subcommittee." On the other hand, coming too early can be just as bad. A good effort can be wasted "if it is too early and other matters are dominating the legislative agenda. We only handle so many things at a time," according to a senior staff person. Also, keep the congressional calendar in mind. While activity in the congressional environment seldom comes to a complete halt, it does vary over the course of the year. A member observed, "There is a much better chance of having an in-depth discussion with me during a recess period, whether in person or on the telephone." This advice applies to meetings with staff members as well.

(The typical congressional calendar is outlined in Appendix G.)

9. Understand Congressional Limitations.

A recurring theme is that too many people bring problems to Congress and "look to us to devise a solution instead of presenting a plan for us to consider, modify and perhaps adopt," said one staff member. It is important to have a good understanding of just what Congress can do and what it cannot do. A committee staff director said, "We don't have big planning staffs that can sit down and spend days analyzing what somebody drops in our lap—such as a ten-page memo with forty-five appendices." Enormous time pressures from multiple competing interests don't leave much time for original analysis and extensive research. Bear this in mind in your contacts with Congress.

10. Make It Easy for Those in Congress to Help You.

State your problem or issue clearly and suggest what action is needed. In describing a meeting with one group of scientists, a senator said, "They were with me for twenty minutes, and when they left I still had no idea why they had come to see me." Avoid this mistake—get the problem or the issue and your request on the table right away. Work carefully at honing your request or advice or information so there is no doubt about your issue, your position, or what you are asking for. Do this by working out a proposed answer to your request or by presenting a plan of actions to accomplish what you desire. Occasionally this might be seen as presumptuous, but more often it will be seen as well organized on your part. Members and staff appreciate proposals for action that are clear and articulate, and show that they have been thought through before presentation. Congress, if it moves on your proposal, may use your language or specific suggestions. Have the material ready to use!

11. Keep the "Bottom Line" in Mind.

In whatever way you are working with Congress, never forget for a moment what your objective is. Make it clear to them as well. If you have a hidden objective or agenda, this is not the book for you. Go back and read Machiavelli's The Prince instead.

12. Use Time—Yours and Theirs—Effectively.

Members and staff are keenly aware of the value of time and resent having it wasted. Plan your efforts in detail and try to make your presentation as concise as possible. Being disorganized or long winded (on the telephone, in writing, or in person) is a sure way to limit your success and future congressional contacts.

A senior staff person cautioned, "You need to remember that staff is generally overworked, is nearly always pressed for time, and generally handles many issues besides the one you are interested in. While interested, they may not have the level of zeal for your project that you have." Do not overload them with details or stacks of paper. It is often useful to have visual ways to make points quickly and effectively. One staff member said, "I look for good ways to brief my boss quickly." A bedrock theme from all staff is that severe pressure on time colors everything they do, including meetings.

13. Remember That Members and Staff Are Mostly Generalists.

While most are "quick studies," you cannot assume that they will immediately understand or appreciate the value of what you are proposing. Do not expect members or staff to have deep familiarity with specific pieces of legislation, or to know their provisions or even their bill numbers. You will lose them if you if you toss out statements like "Section 222 of Title III of H.R. 4494 will kill us." Be concise, but make clear what you are taking about. Keep messages simple, don't be too detailed, and don't overwhelm your listeners with technical jargon.

14. Don't Patronize Either Members or Staff.

Even if it is clear that the person with whom you are dealing is uninformed or misguided, keep your cool and maintain a steady course. Don't resort to an "I'll show this idiot" attitude. On the other hand, it is not necessary to accept rudeness or insulting behavior meekly. While not frequent, instances of such conduct do occur. A call or letter to a member or chairman is one way to respond. Finally, there is always the "Hill grapevine," which can be available through friends, association offices in Washington, and reporters who cover Congress.

15. Don't Underestimate the Role of Staff in Congress.

While it is important to remember that members are elected and staff are not, staff members generally play influential roles in the congressional setting. Do not make the mistake of looking down on a staff member or underestimating his or her ability to help or hinder you, even if the person happens to be very junior.

16. Consider and Offer Appropriate Follow-up.

Seldom will a single meeting with a member be all that is necessary to achieve your objective. Possibilities range from a simple follow-up telephone conversation or two with a staff member to an extended period of working with staff. Conceivably, other members might become involved. Take this into account and be certain that follow-up commitments can be met before you offer them. Before you leave any meeting with a member, try to have clearly identified the name and phone number of the staff person who will be your principal follow-up point of contact. Finally, it is useful and appropriate to ask such a staff member if he or she thinks you should contact other staff members about the issue.

Remember your friends and thank them often. These are more than simple courtesies; they are also the hallmarks of polished professionals. Keep track of your advocates and look for ways to express your appreciation. Use handwritten notes to stay in contact. Private thanks are sometimes appropriate, but also look for public ways to thank them for their contributions.

17. Remember That the Great Majority of Members and Staff Are Intelligent, Hardworking, and Dedicated to Public Service.

If you approach members of Congress with a positive outlook based on the recognition that on the whole they are competent and dedicated, the experience is much more likely to be favorable and fruitful. They need and want your help: make it easier for them to use it effectively.

MEETINGS

SCHEDULED, FORMAL MEETINGS WITH MEMBERS AND STAFF

One of the most frequent ways in which people from outside work with members and staff is through scheduled, formal meetings. Such meetings, which may involve either individuals or groups, are generally held to discuss specific requests, ideas, pieces of legislation, proposals, and the like. The previous section gives you a good start in getting the most out of a meeting with a member or staffer. You should also bear in mind a number of more specific pointers on style and strategy:

Being on time is critical. You are on the member's (or staffer's) turf, and most likely you asked for the meeting. Be aware that the chances are good that you will have to wait when you get there—even for a meeting with a staff member. Much goes on in the congressional environment that is beyond the control of individual members and staff. Although some members and staff are notorious for not holding to their schedules, most make a serious effort not to be late with their appointments. Build the prospect of delay into your schedule; don't take it personally or get upset. Use the time to relax or chat with a staff member who offers conversation. On the other hand, don't interrupt a clearly busy staff person or an overworked receptionist trying to cope with ringing telephones.

Be adaptable and flexible. The congressional environment can be somewhat chaotic at times. In practice, this means that even if your meeting has been scheduled for weeks, it may start late and it is subject to interruption for any of a number of reasons—floor votes, committee votes, or telephone calls from other members on urgent matters. Accept such interruptions gracefully. Don't be flustered by starting and stopping. Think in advance about how to pick up the threads of the conversation and weave them into your next point. Watch and listen carefully to see if you've made the transition successfully. A very quick review of your earlier points may occasionally be necessary, but don't even think of repeating everything.

Even at the risk of some oversimplification, be concise. Don't waste time on too much background. And don't overwhelm them with details; instead, highlight the key facts. If they want more detail, they will ask you for it. In the words of one senator, "Time is of the essence. Make your best case quickly and up front—and let the rest happen." Another senator suggests that you not forget a purpose of mutual value to you both: "Think in terms of providing a basic education about your issue and its broad context to larger congressional concerns. Do this as clearly as possible." Putting the issue in broad context does not mean a lengthy background review. It means tying your issue into one or more of the handful of large concerns facing Congress at any given time—for example, the budget deficit, health care, the state of the economy—and doing it clearly and concisely.

Organize your presentation so as to allow for questions and discussion. One representative said, "I expect to ask questions and I like straightforward answers." A senior staff person advises, "Give a short, clear answer first—and a long answer if the circumstances lead to developing the latter. Only add details and qualifications with encouragement." Another staff member advises: "Be open to all questions even if you think they are stupid or ill-informed—or reflect the views of your opposition." One staff member has some unusual advice: "Consider the members as rather bright, intelligent students who are not terribly well informed on your issue." If you don't know the answer to a question, say so. A staff member says, "Don't pontificate" and don't ever "fake it" with a guess or a confusing nonanswer. Your credibility can sometimes be enhanced by saying "I don't know" if you don't. If pressed, you might speculate and label your response appropriately. Enough people violate this rule to cause members and staff to underscore how strongly they feel about trusting what a person says. Members and staff work in an environment where one's word is one's bond. Do all of them follow this principle? Of course not. But don't misrepresent either your own or your competitor's positions; it will eventually come out. A related point suggested by a number of staff members is "Don't oversell your case." Work hard at building your credibility; it is a tremendous asset—even if your issue is weak or unpopular. To further enhance your credibility, acknowledge as accurately as you can those who disagree with you or are opposed to what you are suggesting; tell the member or staff person as best you can why this is so. Don't make them research this information or be surprised by your opponents.

Be especially careful in planning group presentations. It may be ego-gratifying for every member of a group to have some part in a presentation (or there may even be a valid technical reason for this), but group presentations should be used sparingly, with caution and careful planning. One person must be in charge and manage the individual presentations smoothly but firmly. A staff director advised: "Plan carefully who is going to say what. Don't have a confusing scene in the member's or staff member's office about what is going on. Don't get into side-discussions within the group. And think about this mathematical fact: three people cannot each give a ten-minute presentation in a twenty-minute appointment. This seems obvious, but it is tried often enough to boggle your mind."

It is even more important for a group to have an advance dry run than it is for a single presenter. Use a detailed checklist that the group commits to beforehand. One long-winded presenter can ruin your entire show by forcing a carefully crafted closing of five minutes to be done in one-minute—with the resulting loss of focus and force in achieving your objective.

Staff members can be powerful and influential, often serving as gatekeepers who can help you communicate with members. Don't underestimate their value by thinking you have been passed off to an underling if a member did not agree to your request for a meeting. In the words of a committee staff director, "Staff members have a lot of discretion on who sees members, on who testifies at hearings, on what is read by members, and on what goes into legislation in the form of specific words and sections." In setting up meetings with members, it is often useful to enlist the aid of a staff member before approaching the member.

Remember, though, that not all staff members are equal. This is a basic fact of life in Congress. As part of your intelligence gathering, assess the roles of those staff members who appear to be important to you and use the information effectively. Include in this assessment member-staff relationships. Some staff members have wide latitude in speaking for members; others aren't authorized to tell you the time of day. Some staff members can directly influence multimillion- or multibillion-dollar programs; some simply gather information for others to use. A senior staff person advised: "Make sure you are talking with a person who can really do you some good." However, a companion to this advice is not to be upset or put out if you end up with a less than influential person; there is always another day and people do get promoted! Finally, as another senior staff person suggests: "Don't confuse staff interest in a meeting on your project or issue with automatically leading to a favorable outcome for you."

While a personal staff member can be the conduit for you to see a member, and he or she can serve as a liaison to appropriate committee staff members, a personal staffer will less often be the focus of major legislative or budgetary activity, especially in the House. Committee staff are likely to be more legislation oriented or technically informed than are personal staff. Personal staff are likely to be more member oriented and district or state oriented. All of this means you must do different kinds of homework and plan your meetings accordingly.

Informal Visits

Not all circumstances lend themselves to formal, scheduled meetings. Sometimes it is useful to visit with a member or a staff member for purposes more limited and less formal than a presentation or a request for action or help. Occasionally—and on a selective basis—brief "stop-by" or "pop-in" visits may be used. Either with an advance telephone call or on an impromptu basis, your visit should be advertised as a short, one-minute-or-so affair. Such visits can be for a simple "Hello—I'm in town and hadn't seen you in some time," or, for example, passing along some favorable or valuable news, or to provide an advance copy of an important report, or for saying that a report has come out or is coming out and asking, "Would you like a copy?" However, stop-bys or pop-ins are seldom, if ever, to be used unless you have some relationship with a staff member who may place value on the connection you represent. So, be careful about getting yourself into an awkward or embarrassing situation. For a member, unless you are a visiting constituent or have a good personal connection (e.g., a role in a campaign), forget pop-ins, at least in Washington. While many of the points discussed earlier apply to informal visits, there are also some additional items worth highlighting.

Informal visits can play an important role in building and maintaining good working relationships in Congress. Visits can help you avoid the mistake of not being seen until you have a problem and need help. (This doesn't mean that first-time visitors or requests will not receive a fair hearing. Negative comments by members and staff about those who are "invisible until they need help" tend to be aimed at organizations and individuals who make a practice of this.)

Overall, the matter of informal visits calls for finesse and good judgment. On the one hand, building a relationship is important, but on the other, you certainly don't want to be seen as a pain in the neck who hangs around an office wasting time or who stops by to kill time while waiting for an appointment elsewhere. This kind of behavior can alienate staff members and can virtually ensure that if you ever need help, it will be difficult to get it.

While informal visits need to be carefully considered in Washington, they are easier to pull off in the district and state offices. Often this is your best opportunity to meet with a member and command a decent attention span. Members and staff observed that "too many people don't understand that members can be seen in their district and state offices." As noted by one senior staff member, "This is where you can walk in and see my boss without an appointment. A lot of members hold 'open-office' sessions for their constituents." Also, personal staff members from Washington visit the district and state offices from time to time and may well be available there. Not to be forgotten, however, is the overwhelming orientation of district and state offices: constituent service. (See the last section of this chapter for more advice on working with district and state offices.)

Constituent service is an important part of congressional life. As a U.S. citizen, you are a constituent to one representative and two senators (unless you live in Washington, DC, or in the U.S. territories). In addition to a visit to a district or state office, you might consider visits to the Washington offices of your members if, you are in town on business or even on a holiday visit. Even in this case, though, a phone call ahead of time is simply good manners. You may or may not get to see your members, depending on their schedules, but are likely to get to see a staff member. If you are in Washington for a hearing, provide your member with a copy of your written testimony—preferably with a summary. It is not uncommon for a member to come to a hearing and introduce a constituent, however, don't count on this. It will depend on a number of things, including the member's interest in your topic, your status, and the member's schedule.

MEETING ADVICE WRAP-UP

On any given day, members and their staffs will be talking with many different people on many different subjects in all kinds of meetings. You are competing for the member's time and attention. However, knowledgeable people who present their message with clarity, who make their request or offer simply and concisely, and who generally make it easy for the member or staff member to help them are such a rarity that they will be remembered, helped if at all possible, and called upon in the future.

USING THE TELEPHONE, FAX, AND E-MAIL EFFECTIVELY

Use of the telephone dominates the congressional environment. In the fast-paced, verbal setting of Congress, this is one of the preferred modes of communication for both members and staff. Using the telephone skillfully can greatly advance your relationship with staff members.

In the survey conducted for this guide, dozens of staff members mentioned their receptivity to telephone contact. Staff described themselves as "always open," "frequently open," "very open—unfortunately for me," "yes, I'm here to serve the public," "as time permits," "generally open—once the relationship is established," and "will usually talk with about anyone, but realistically I'm more open to calls from people or groups known to me." One staff person said, "I return all of my calls eventually, but I acknowledge that not all staff members do so—particularly if they don't know the caller."

The ready availability of staff should be no surprise. However, many people do call for a member and are upset when it is not possible to be put through or when they are referred to a staff member. Don't be put off if you are referred to a staff member. Understanding the reality of working with Congress involves a recognition of the special role played by staff. This applies to telephone calls as well as meetings. Also, within the staff hierarchies it is more difficult to reach some staff members than others.

Even for a telephone call, and even with a junior staff member, prepare a little checklist. It truly upsets congressional people to get into a telephone conversation with someone who, as one staff person described it, "wanders all over the barnyard looking for the chicken." Making the checklist will help you to avoid the problems of wandering around or of conveying confusion because you're not sure you covered everything you intended.

The first order of business is to get your name, organization, and purpose out on the table. If you have a personal connection with the individual you are asking for, use it. Very often, you will be going through a busy receptionist, so have this information ready. If you are put through, or the person answers directly, make sure, as one staff person suggested, "that it is not a bad time and say that you would like perhaps five minutes or so." Another added, "Be sensitive to the staff person's schedule—don't call the morning of a hearing to discuss philosophy!"

Stating your business clearly and quickly is essential. One staff member put it this way: "Remove the wonder of why this person is calling me. Don't make me guess at your motivation or why you are calling." Another staff person advised, "Don't obscure your message because you're embarrassed about asking for something. Come right out with it." Still another said, "Unless you are a personal friend, don't waste time with chitchat." Others said: "Please, no long backgrounds. Get to the point. Know what you want." "In addition to many visitors, I have 50 to 100 telephone conversations a day. I don't have any time for long-winded baloney."

As much as possible, do your homework before making a call, although it is quite acceptable to call someone for advice on just who is the right person to talk with. However, do not waste the time of a busy staff person asking for routine information on legislation or other matters that is readily available in databases or published sources.

Avoid using telephone calls for complicated subjects. Not every subject or problem is appropriate for handling on the telephone. Typical staff member advice was "if the issue is complicated, make an appointment. Don't even try to use the telephone." Still others said they preferred a letter in advance of a telephone conversation. An admonition was to allow a reasonable time for the letter to be read. Still others suggested they preferred a letter in advance if they did not know the person.

Be patient, but persistent. Remember that members and staff are under horrendous time pressures. As one staff person described it, "We may not be able to return calls immediately or even the same day. And some staff return calls late in the day, so you may want to consider leaving a home number." Others observed there was nothing wrong with being persistent if you have not heard back after several days. Just try again, with politeness and aplomb.

Remember that telephone calls based on "outrage" and "demands" don't go over well. As a citizen you can exercise your First Amendment rights to call anyone in government and tell them how rotten you think they are running things.

However, such calls seldom lead to anything. Several staff members pointed out that "Rudeness doesn't help. And I don't respond very well to demands that I do something."

Depending on the nature of your telephone conversation, you may wish to consider a follow-up note in a letter or a fax. It may be a simple "thank you" or as one staff member suggested, a summary of the details of the issues or discussion. Faxing is often a useful alternative to playing telephone tag. All congressional offices have fax machines, and—depending on the issue and personalities involved—a brief fax message can sometimes gain a response more quickly and efficiently than is possible by telephone. Bear in mind, however, that congressional offices can be overwhelmed by incoming fax traffic, just as they can be swamped by telephone calls. Use this mode of communication judiciously.

Use of electronic mail is also growing on Capitol Hill. As of December 1995, the "Thomas" World Wide Web site at the Library of Congress (http://thomas.loc.gov) listed e-mail addresses for 140 representatives and 64 senators. In addition, a number of committees (including the House Science Committee), leadership, and other offices participate in the congressional electronic mail system. The directory on the website emphasizes that the system is primarily intended to provide a vehicle for members to communicate with their constituents, suggesting that other persons wishing to communicate with members use alternative modes. This is perhaps not surprising, in view of how easy it would be for a grass-roots lobbying effort to flood members' mailboxes with junk e-mail.

Since this technology is evolving at a rapid rate, you should probably seek current information before trying to use it for the first time. The House and Senate also have gophers, and the House has an e-mail help line at househlp@hr.house.gov. Don't be disappointed, by the way, if you receive an automatic reply or a reply from a staff member rather than the member to whom your message was addressed. Few members have the time to respond personally to more than a few of the e-mail messages they receive.

TELEPHONE, FAX, AND E-MAIL ADVICE WRAP-UP

While Congress is extremely open to telephone calls, don't abuse this openness with trivial and poorly planned calls. Prepare in advance with a checklist. Identify yourself—with a referral if possible—and get to the crux of your business promptly. Use the same kind of planning for your fax and e-mail communications as you would for meetings and telephone calls. Don't fall into the trap of sloppy use of faxes and e-mail. Never be rude or demanding; politeness, patience, and persistence will pay off.

PREPARING AND SUBMITTING CORRESPONDENCE

Although use of the telephone is one of the defining characteristics of Congress, letters by mail or fax are a common feature of congressional business. You may be sending a letter to ask for a meeting, to ask for a telephone conversation, to describe a problem that calls for some action, or to provide information. Do not let the ease of faxing beguile you into careless planning and preparation. Maintain the same high standards in a fax (or e-mail) as you would in a traditional letter.

Among the most useful letters and reports are those that provide information in easy, readable, and understandable form. As one staff member observed, "I find them useful if they contain constructive comments on pending legislation, especially if they represent consensus positions taken by reputable groups." Another added, "they are valuable to me when they point toward issues of mismanagement, etc., that are not widely known already." Oversight and investigative hearings have actually resulted from letters of this type. "To be effective, however, letters must be written with care," said one staffer. "I hate trying to wade through prose that sticks to your brain cells like mud on your shoes." A staff director suggested that the most useful letters include information not easily available otherwise. Several staff members underscored that "good anecdotes" are always helpful.

Brevity is a word that comes through often in talking with members and staff about correspondence. As one staff member said: "Volume kills! An A-Number One letter for me is one page." A clear, concise, well-reasoned presentation of an issue, problem, or request is what is desired. Getting to the bottom line quickly was the advice of one senator: "Tell me what you want or have to say in a nutshell." On this point of focusing your letter, a senior committee chairman said, "Put the details in an enclosure." A senior staff person advised: "Don't be coy or beat around the bush. Get to the point. Be up front about what you want." Do not clutter your message with an overabundance of facts. Keep them short, selective, and pertinent.

On the other hand, don't assume familiarity with your issue. This admonition of not assuming familiarity may seem at odds with the advice on brevity and conciseness. It isn't. It simply means you must strip down your case and your data to what is absolutely necessary to communicate your main objective. Your letter should be a model of simple elegance, in which the reader can flow easily through your prose despite not being the expert you are.

In your letter, demonstrate that you have done your homework and know how your issue fits in. Make clear that you are aware of various views on an issue. As long as you are fair and accurate with your facts, there is nothing wrong, as one staff person said, with "telling us why the others guys are wrong."

It is not always clear to a member or staff person why they should become involved with you or your issue, so part of your letter must be devoted to making this case concisely. It is also important for you, as advised by a senior staff person, "to describe what else has been done—that is, what you have done to help yourself, other than to contact Congress."

Tie your letter to a member's district or state interest. One Senator suggested that the letters that get read are the ones from his state. "From a crass point of view," said one staffer, "it helps if the letter comes from the chairman's district." Some in Congress particularly like letters conveying political information—such as "I'm upset about this particular piece of legislation..." or "This specific program is a problem because..."

A little personal information is necessary if you are unknown to the proposed recipient. Clearly indicate your addresses, telephone numbers, and times you are available, one staff member noted, "It's surprising how small things like this are not attended to properly. It makes it much more difficult to follow up."

Two Cautions. Be wary of using mass mailings or form letters. A good way to raise the blood pressure of a staff member and even members who are used to this sort of thing is to mention mass mailings. Various staff people said that they ignore mass mailings and form letters saying: "I've been a victim of mass mailings and I don't like them." "I don't pay much attention to `tear out card' mass mailings." Still, while orchestrated lobbying efforts through mass mailings are generally discounted, they have worked in some circumstances, according to various staff members. It usually takes a seasoned lobbyist to carry out such an effort successfully, however, so approach the concept with care.

Finally, don't seek to become pen pals with staff or members. Some staff and members will shudder and show voluminous files from "pen pals" who spend much of their time writing letters to the office, or so it seems. Don't take a friendly response to your letter as an invitation to undertake an ongoing correspondence.

CORRESPONDENCE ADVICE WRAP-UP

Remember your goal in preparing the correspondence: to get it read and acted upon. Make it easy to read. Organize your material with care. Work to make it concise and clear. And test it on others before sending it. Individualized letters probably have the best chance of being read—as compared to a form letter or tear-off card. Letters that are polite and politically realistic are better received than polemics and are more likely to join the select stack of letters that go to a member for a personal look and a response to a staff recommendation.

WORKING WITH STATE AND DISTRICT OFFICES

State and district offices offer a relatively easy way of gaining access to members who may be difficult to see in Washington. However, all too few scientists and engineers seem to be aware of this channel.

Many representatives and senators have walk-in appointment periods in their state and district offices—although this does not mean you should just walk in without any preparation. Find out where these offices are located, the schedule for such periods, and the procedure your member follows to use the time slots. You can obtain locations and telephone numbers through your local library or telephone directory, or by calling your member's Washington office.

Most congressional offices place legislative and budget staff members in Washington. The state and district offices are more likely to be staffed to handle a variety of constituent services. Yet this does not mean that you cannot discuss your issue with a local staff member. For example, you can get advice on just whom to contact in Washington or on the feasibility of meeting with the member at a future date. However, unless your issue is "hot" locally, you could well be referred to the issue person in Washington. Even so, the district and state staff will usually make every effort to be helpful.

Meeting with the member in the district may be more relaxed, but that doesn't mean you should be any less prepared than you would be for a meeting in Washington. Focus on making it a memorable and valuable experience for the member as well as for yourself. Even in the district office, you must "rise above the clutter." This includes taking advantage of the opportunity to relate your concerns to local issues and to give examples of how your colleagues feel about the issue you are addressing—if you know those facts. Tying in other constituent interest gives support to your position.

The time your member spends in the district gives you a special opportunity to show science and technology at work by arranging visits to local laboratories, colleges and universities, events at local scientific societies, or a company facility. This gives the member a valuable window to how scientists and engineers operate and interact and provides a more informal opportunity to convey the importance of policy and budget decisions in scientific areas. You can work with the local offices in requesting the participation of your members in local chapter or regional meetings of your societies. If you give enough notice and have some flexibility in your request, members say that the chances are good that they can appear. As a practical consideration, do your best to assure a good turnout. Spending an evening talking to five or six people may not be the most effective use of a member's—or a staff member's—time. For example, one senior staff member reported: "My boss had to cancel out of a district event at the last minute because of a late floor vote and I went in his place. They had told us there would be a hundred people, but due to faulty publicity, it ended up with less than a dozen. Fortunately, I did some other district business so it was not a total waste, but I sure wasn't very happy with that group."

Members say they are interested in visiting district organizations where interesting work is being done or where they have an opportunity to talk with workers. As one representative said: "I like to know when a company in my district has come up with some new product or development that is significant." Another representative noted that "Over the years, we have developed good relationships with a number of researchers at universities in my district." He identified these relationships as among the most favorable of his experiences in working with the scientific and engineering communities. You, too, can work at developing such relationships: your work gets exposure to the member, and the member gets information as well as exposure to your colleagues. Building such a relationship can eventually result in your serving as an informal advisor, providing information, opinions, and perhaps more formal studies on matters of importance to the member.

Apart from your working with a member of Congress and his or her staff as a constituent seeking assistance or giving advice and information, there is the special role of serving in a political campaign. This is a difficult issue to raise, but it deserves to be placed on the table. Scientific and technical issues are political in the same sense that other areas of our modern society are political. Decisions entail setting priorities and allocating budgets, and these are quintessentially political decisions. Members of Congress have been strong supporters of science and technology but have not seen this translated into political support in the same way that being a proponent of labor or business or agriculture generates support.

By no means does seeking assistance from a member of Congress translate to an obligation for political support on your part. However, there is a perception—accurate or not—in the world of politics that a sizeable number of scientists and engineers believe they are "above it all" and that being involved with politicians is inconsistent with the ethos of the scientific and engineering professions. While many interest groups provide support—financial and otherwise—for increasingly costly political campaigns, various members of Congress claim to have observed a general aloofness on the part of the scientific and engineering communities.

Obviously, support for science and technology should not be the sole basis on which you make your political choice in an election campaign. But if there is a general compatibility on issues and you believe the individual has been doing a creditable job in office, then you should think about taking part in the member's campaign. This can include arranging voter meetings, organizing fundraising events, publicizing member achievements and contributions, and making financial contributions. This is not the place to describe in detail the operations and functions necessary to an election campaign, and your decision will depend on your own personal style, abilities, and interests, but the point should be clear: politics is important to science and engineering. Consider getting involved.

5 | HEARINGS AND TESTIMONY

Chapter four provided guidelines on meetings, visits, telephone calls and correspondence with members of Congress and their staffs. In one sense, these activities can be considered as preliminaries to performing in the center ring of the circus—appearing as a witness in a hearing before a congressional committee. While the reality may be that a private meeting with a powerful member or staff member can be more influential than testifying in a public hearing, there is a strong belief—inside and outside of Congress-that holding public hearings is one of the more important things that Congress does. In recent years, Congress has averaged upwards of four thousand committee and subcommittee hearing per year, calling upon tens of thousands of witnesses.

Congress spends a lot of time and energy on public hearings. Yet, with thousands of witnesses appearing annually—perhaps a dozen or more at any one hearing—there is a distinct and difficult challenge for any witness. As one senior staff member described, the challenge is "to rise above the clutter." "We are deluged with paper and telephone calls, hearings, meetings, and visits," this staffer said. "You have a lot of competition, and you have to figure out a way to distinguish yourself and what you have to say from all the thousands of others who are likewise competing for our attention."

There is relatively little published material explicitly devoted to testifying before Congress, but probably the best single publication is *Testifying with Impact* by Arch Lustberg. Lustberg's booklet covers virtually all aspects of testifying, from breathing exercises to relax yourself to avoiding "deadly" facial expressions and other mannerisms to general advice on communicating more effectively.

> "We take a lot of time to understand science; please take some time to understand us. Keep in mind you are dealing with focused generalists, not narrow specialists; get to the point in an understandable manner."
>
> Rep. Sherwood Boehlert (R-NY)

PREPARING WRITTEN TESTIMONY

First, understand the purpose of the hearings. Regrettably, not all witnesses come to a hearing with an adequate understanding of just why the committee is holding it. In practice, this also means that you should answer—or attempt to answer—the questions asked in your invitation letter or telephone call. Staff members observed that many witnesses don't do this. One firm admonition was to "be responsive to the subject or purpose of the hearing." Take the opportunity to make your own pitch, but first you should respond to the needs of the committee and its invitation. Earlier advice on intelligence gathering applies here as well: find out "what really is going on."

Consult and work with committee staff members as far in advance of the hearing as possible. You can learn much about how best to focus your remarks on issues raised by the committee, get a better understanding of the turf and the personalities, and pick up general advice on how to present yourself most effectively. Unless you are appearing under a subpoena (a special circumstance not covered in this guide), consider the committee staff as allies who can ease the way for you. Sometimes they will tell you in advance what questions their bosses are likely to ask—because they often draft these questions. Often, committee staff members will work with you to develop questions that might help clarify an issue.

Both your written and your oral testimony will become part of the written record. Prepare your remarks with simplicity, brevity, and clarity. These are admonitions that run through all aspects of working with Congress, but they must be emphasized in preparing testimony. A staff member advises, "Your audience doesn't have a scientific or technical background, so write for the layperson." Another said, "Make it understandable for the congressperson, with concrete examples if possible." As a specific suggestion, a member said, "I like to see a bullet approach with tightly written phrases. This helps me to understand more quickly what you're getting at." A senator said he found it helpful for a statement "to use imagery that can capture my attention." Still others observed that, on the whole, members are uncomfortable with abstractions. Another senator implored, "Write concisely; give me clear, articulate analysis; put the details in attachments."

Your main message should come through early. Put the details or backup data in appendices. While it is acceptable to provide a background summary, don't overdo it. And make sure your vision comes through clearly. As a staff member explained, the committee must be able to understand "where things stand; what needs improving or changing" from your testimony. One senator called for a statement to highlight the most compelling data and to have a distinct "bottom-line orientation." Another staff member said, "We like statements that convey facts, contain original analysis, and clearly state a position." Finally, a committee chairman said, "I want to understand what course of action you are proposing and your justification for it."

> "I always find it helpful if a group seeking help from Congress outlines what they are doing first to help themselves. Too often we see an entitlement mentality at work among those who lobby us for funding."
>
> Sen. Jeff Bingaman (D-NM)

In presenting your case, it is quite acceptable to be candid and forthright—as long as you are sure of your facts. A senior staff member put it this way, "You can be frank—but factual." It can be worse than just embarrassing if you are found to have been sloppy with your facts. If your situation calls for the use of information that needs to be qualified, do so; where appropriate, characterize as best you can the nature or range of uncertainty. As one staff member explained, "You're here as an expert witness in the court sense, not as a truth-seeking scientist, and it's OK to advocate a point of view."

There is a balance to be achieved here: congressional testimony is not a scholarly document, but it should be well documented. Avoid voluminous footnotes; brief citations are in order. Also, statistics, graphs, and charts can be desirable if they make clear points and are used with discretion.

Think carefully about how best to present your testimony. Play with the sequence of your facts, suggestions, observations, and conclusions to find the most effective way to make your case. One good way is to begin with a stripped-down outline of no more than one page. Organize and reorganize this outline until it comes together.

Check in advance on the desired format of the statement and comply with committee administrative requirements.* A number of items can, collectively, make or break your testimony, or at least strongly influence the way it is received. Submit your written statement with the required number of copies by the deadline requested by the committee; this will vary, but it is often forty-eight hours in advance of a hearing. Unless you were literally asked at the last minute, bringing your statement with you on the day of the hearing is not acceptable. Submitting testimony in advance of the hearing enables staff (and sometimes members) to review your statement and prepare questions for the hearing. This greatly enhances your opportunities for dialogue with committee members.

> "It's important that you make clear what your priorities are. Nothing weakens a presentation more than giving the sense that a particular item is the most important thing you want—except for all the other items that are also the most important things you want."
>
> Rep. Robert S. Walker (R-PA)

---

* Some committees require single-spaced testimony of no more than five pages; others aren't so specific. In general, however, double-spacing with ample margins and space between sections should be used as many members and staff like to mark or make notes on a statement.

PRESENTING ORAL TESTIMONY

It may be helpful to think of a hearing as theater. One senior staff member provided a perceptive insight into hearings by observing that although testimony is not a speech, it must still be listened to by an audience. In his experience, it was most like theater with a specialized audience. Thus, his advice was to consider presenting a statement as a "theatrical performance" that you as the witness/actor ought to prepare for just as you might for an opening night. In developing the imagery, he said, "Don't make it difficult for the audience to endure. There have been times when I wanted to jump up and shout, 'Shut up and get out of here.' There have been other times when I wanted to applaud." And the sad part is, he noted, "that some witnesses are so bad they don't get listened to even though their message may be very important." Finally, one senator said, "I like an enthusiastic witness—and one who gives me concrete examples."

"Since everyone in Congress is pressed for time, be direct and concise. Do not go into a meeting without having a clear idea of your purpose and the main message you want to convey."

Rep. Rick Boucher (D-VA)

Make sure you are thoroughly familiar with your statement. This cautionary note is offered primarily to those who may be delivering testimony that has been prepared by others. By the time you present the statement, it should be yours. Go through the statement line by line to make sure there is not some fact or conclusion with which you are insufficiently familiar. Know the main points well enough to present them without reading a single word if that becomes necessary.

With few exceptions, you can expect a committee chair to announce that "without objection, your full statement will be made a part of the hearing record." You will then be given a certain amount of time—usually about five to ten minutes—to summarize your statement. The time allocated to you may be even less if the hearing is running late, as often happens. If there have been interruptions for members to leave for a vote on the House or Senate floor, you should prepare yourself for a request to make your summary very short.

Your summary statement should be crafted and presented so as to set the stage for a good question and answer session with the committee members. As one senior staff person put it, "The idea is to get a dialogue with the members while you are there. They can read your statement while you are elsewhere if you grab their interest."

You will seldom have any choice about whether or not to summarize your statement. In general, however, you will have some choice about how to do this. Work with committee staff in making your decision on presentation style and which of the following models to select. But above all else, do not "just wing it." NOTE: It usually takes two to two and one-half minutes to read one double-spaced page. This means a maximum of four to five minutes for a ten-minute summary. Some approaches to consider in preparing your oral presentation include the following:

> "In general, we trust information from scientists. But to keep that trust you must clearly state which of your points is opinion, theory, or widely accepted fact."
>
> Sen. Pete V. Domenici (R-NM)

- Mark your prepared statement to highlight the way you will summarize it. Include "road-map" instructions to let the committee know where you are in your prepared remarks as you go through your summary.

- Prepare a separate, shorter version of your statement to be given orally in summarized form. Give copies of this short version to the committee staff when you arrive at the hearing for distribution just before you present your testimony. This approach serves the needs of those members who are more print oriented or who may have to miss part of your oral testimony.

- Use a short version of your statement in presenting your testimony, but do not give copies to the committee. Make clear, however, that you are not using your prepared written statement. This will minimize the tendency of members to leaf through their copies of your statement attempting to see where you are.

■ Use a short, written one-page outline of your key talking points as a road-map to guide your presentation, making it seem extemporaneous. Practice this routine a number of times so as to present eloquently within your time limit. This approach and the previous ones are designed to keep you from wandering off track—as do so many witnesses.

Be careful about going off on tangents. Think of your presentation this way: You have ten minutes to make five points, if you spend five minutes on your first one because you have gone off on a tangent, you're in trouble. According to members and staff, this is the error most frequently committed by witnesses, especially those who attempt to summarize without good planning.

> "'Compromise' is a dirty word to a scientist. To an engineer it means trading off conflicting requirements. In politics, it is the only way to get anything done."
>
> Sen. John Glenn (D-OH)

But even experienced witnesses can fall into this trap as an interesting idea occurs to them and they pursue it. All too often, witnesses get into the embarrassing situation of having used up their time to cover only half or less of what they had wanted to say. Strict discipline, careful preparation, and practice are essential to avoiding this serious problem. As one staff person said, "Make your points briefly and then say 'I'm open to questions.'"

It is often helpful to find out in advance who else will be testifying and what their key points will be. You will then be better prepared if you are asked during the hearing to respond to positions taken by these individuals. Although you can disagree politely, don't be overly critical of the testimony of others. Remember, hearings often are deliberately set up to hear opposing points of views.

Arrive well in advance of your scheduled appearance. If you are the lead-off witness for a hearing, plan to arrive at least thirty minutes in advance of the scheduled start. Arriving early could afford you a few minutes with a committee staff member with whom you may only have talked with on the telephone. It is even possible that you may obtain some last-minute intelligence on the hearing that could be of value in your presentation.

> "Too often presentations are unfocused; there is not enough information on what is the problem, what is the proposed solution, and what decision is required."
>
> Rep. George E. Brown, Jr. (D-CA)

Even if you are not the first witness, it is still important to arrive early and listen to the witnesses preceding you. You can get a sense of how the hearing is unfolding, what kinds of questions are being asked, and which members are active in the process. It is impressive to the committee when, later, you can refer knowledgeably in your own presentation to some earlier point made by another witness or, better yet, a member.

In *Testifying With Impact*, Lustberg describes the skills and techniques of an effective presenter and speaker. Reinforcing Lustberg's guidance, staff and members responding to our survey made such suggestions as:

- Speak clearly.

- Avoid looking down at the table; seek eye contact with members. One senator said, "Talk to me."

- Be direct and assertive but not overbearing; convey an air of confidence about yourself and what you are saying.

- Be animated. It is the kiss of death to sit and read a statement in a monotone. "Speak from the heart," said a senior staff person, "and don't worry about your grammar." Try to convey your message with excitement, enthusiasm and liveliness. Another senior staff member said: "Members like a knowledgeable, enthusiastic witness."

- Choose a style most suitable for the circumstances. Some hearings take place more informally than do others. Some chairpersons are relaxed and easygoing, others are sticklers for procedure and protocol. Inquire about the style of the member chairing the hearing and plan your appearance accordingly.

■ Tell a story, use examples and imagery, and strike a balance: don't be too technical but don't talk down to your audience either. Avoid jargon. Be relevant. Make analogies to other important issues that—based on your intelligence gathering—might be important or familiar to the members and the chairperson.

■ Talk to the committee—not the audience. Liberate yourself from the printed page.

While the name of the game is to "rise above the clutter," it can be dangerous and counterproductive to use unusual or shocking attention getters, particularly without discussing them with staff in advance. If you are going to pass around some item for inspection, be sure it is on point and does not distract the committee from your main message. Gimmicks should be avoided; caution and finesse are in order. This does not mean that controversy is inappropriate or that you cannot be provocative under the right circumstances.

Jokes fall into the gimmick category and should be used only with extreme care and finesse. Quips or putting a humorous twist on some event or comment during the hearing are fine if they fit and if they are spontaneous. However, unless you are good at this, it is better to stick to your testimony and avoid the comedian role.

> "Be succinct, be brief, and be confident. As scientists and engineers you have the technical expertise and knowledge critical to the policy debates here in Congress."
>
> Sen. Tom Harkin (D-IA)

In your prepared statement you can use such charts and figures as you think necessary. In your oral summary, however, you must be much more selective, if you use any at all. "Your time is short, so think carefully about what you might gain by the use of such techniques," reminded a staff member. Think, too, about the logistics of such items, including who will show them and room lighting. Charts and graphs must be well designed and tied into your main points. Another staff member observed that "busy charts are guaranteed to lose your audience." Such devices can serve a useful purpose and be very effective in conveying a point, but make sure they are well designed before you decide to use them. Ask in advance if it is appropriate to bring board charts, slides, or transparencies with you.

In answering questions, remember the following:

- Anticipate questions and prepare for them. In particular, work with committee staff and personal staff to identify potential areas of interest on the part of members.

- Answer questions concisely and directly. As with your oral summary, don't go off on tangents. If members want more, let them follow up by asking for it.

- Draw on your written statement as well as your experience, to reinforce your points or to make additional ones.

- If you don't know the answer to a question, say so. If pressed, give qualifiers or ranges of uncertainty with whatever response you decide to make. Offer to provide the answer for the record, in writing, if you would like to do so.

- It is acceptable to equivocate at times with a "yes, but..." answer, but don't overdo it. As one senator put it after listening to several "on the one hand...and on the other" statements, "We need more one-armed scientists."

- In a panel setting, you may be asked a question by name or it may come to the panel at large. In the latter case, let good manners and common sense prevail. Neither a question hog nor a shrinking violet should you be. If another witness is tending to dominate the responses and you really do have something to say, assert yourself. If a question is addressed to the entire panel, it is sometimes helpful to glance at your fellow panelists and quickly gauge who is most anxious to answer first.

- Don't lose your cool. You can differ politely with another witness or even a member, but do it with grace and good humor.

AFTER THE HEARING

The hearing is not really over when the chairperson gavels the committee into adjournment after the last witness has finished. For the staff, there is still the hearing record to complete and possibly a report to be prepared along with associated legislation. You, as one of the witnesses, will be involved in the post-hearing process.

Respond promptly to questions submitted for the record. It is customary in many congressional hearings for individual members and the staff (on behalf of the committee) to submit questions in writing to a witness for an answer later. While it is more common for such questions to be submitted to organizations, individual witnesses also get them. Think of this as an opportunity to expand on your written statement or your presentation at the hearing. It is possible to negotiate downward the number of questions and the length and nature of the response requested if the request seems onerous. On the whole, committee staff are quite cooperative in this respect and understand the limits of individuals. The main point is for you to respond promptly on those questions that you choose to answer. You may also take this opportunity to offer additional comments on questions posed to you earlier in the hearing itself. Further, you may even suggest questions to the committee staff to ask of you or others.

After the hearing, you will be sent an excerpt of the transcript of the hearing record. This will contain both your oral statement and any questions-and-answer exchanges that took place on the day of the hearing. You will be given the opportunity to correct any mistakes made in the transcription process. Although the details vary by committee, in general you will not be permitted to rewrite your testimony or to fix grammatical mistakes to make your remarks look better. As in responding to questions for the record, it is important for you to respond quickly to the request by the committee.

After the hearing, contact the committee staff for a posthearing critique. Even if only by telephone, you should make a follow-up contact to determine how the staff thinks the hearing went. It would be especially valuable for you to hear how your presentation and appearance were received. This type of critique will not always be possible to get, but many staff are inclined to be helpful in this regard.

The hearing is only one part of the process, and it will be important to follow up to see what, if anything, happens as a result of it. Depending on the outcome, you may want to consider additional actions in contacting Congress.

6 | A CAUTIONARY NOTE

This guide is intended to encourage you to become more actively involved in congressional affairs and to make you more effective in working with Congress. It should help you understand how Congress operates and give you a relative advantage over those who might attempt to become involved without a basic understanding of the institution, its organization, and its traditions. However, neither this book nor any other will make you an instant expert on Congress. We hope that we have merely whetted your appetite. If you have a serious interest in working with Congress, please recognize that this is not a stand-alone source of information. The final piece of advice we have to offer, therefore, is to recognize the limitations of your knowledge—that is, know what you don't know and when to get help.

Bear in mind that the overview in chapters two and three presents a highly simplified picture of Congress. Dozens of scholarly books based on years of research, as well as numerous personal memoirs based on years of experience have been written about Congress. A few of these are referenced in the notes to chapters two and three or are listed among the suggested readings in the appendix. The richness and variety of the many dimensions of Congress as described by congressional scholars, along with the many revealing insights presented in the personal memoirs, are difficult to capture in a brief overview such as this. Couple your use of this guide with additional readings or, better yet, with help from others who are knowledgeable and experienced in working on the Hill. The best guide is a human one, such as a colleague who has firsthand experience in working with Congress and who can give you advice and assistance in the context of your specific situation and needs. Technically trained individuals who have worked in Congress for a substantial period of time—such as the Congressional Science and Engineering Fellows sponsored by many scientific and engineering societies or the government relations professionals in your organization or professional society—are ideal sources of such advice.*

A second caveat arises from the nature of this book, in which experience and guidance from a wide range of individuals has been filtered through the lens of the author's judgment. Not everyone may agree with the advice and opinions offered in chapters four and five. It is important to remember that the "rules" in these chapters are subject to interpretation, to exceptions depending on circumstances, and (occasionally) to out-and-out disagreement. They are guidelines and suggestions to be used judiciously, not followed blindly.

A final note on political involvement—which is both a caution and an exhortation—is in order: In the concluding section of chapter four it was suggested that scientists and engineers who are seriously interested in working with Congress might consider volunteer work in a member's political campaign. Among those who reviewed the manuscript for this guide, there were some rather strong differences of opinion about whether it was appropriate to include this topic.

In the author's view, these differences reflect decades of ambivalence and differences within the scientific and engineering communities about the degree and nature of their relationships with the world of politics. The debates over

---

* AAAS sponsors two Fellows each year, coordinates the Fellowship programs of nearly two dozen other scientific and engineering societies, and maintains a network of former Fellows. For further information contact the Directorate of Science and Policy Programs, AAAS, 1200 New York Avenue, NW, Washington, DC 20005, telephone 202-326-6600, science_policy@aaas.org.

these relationships are now rather more muted than they were in earlier decades—particularly during the 1930s and 1940s—when they were fierce, public, and acrimonious.

The central concern of many scientists in that earlier period was the possible loss of independence on the part of science if it became too closely connected with politics and too dependent on government support. While this is not the place for a thorough discussion of this complex topic, in the author's view, it seems fair to say that, notwithstanding occasional forays by political actors into science in the form of priority setting—especially in the biomedical fields—the scientific community has retained a great deal of autonomy in the allocation of research support within disciplines and in the management and control of basic scientific research.

Although this is not widely appreciated in the scientific community, this independence and autonomy have been maintained as often as not through alliances among key politicians and scientific and technical leaders. Over the years there are numerous examples of congressional figures fending off attacks on science from other politicians and sectors of the public. While there are many reasons for scientists and engineers to work with Congress, self-interest alone is important enough to warrant involvement. All too often we hear scientists and engineers bemoaning the lack of scientific and technical understanding in Congress. If we, as scientists and engineers, expect Congress to understand us, it is essential that we make more of an effort to understand and work with them.

Glossary*

APPROPRIATIONS: Money legally set aside for a specific use. Appropriations are determined annually by Congress for the following fiscal year for every program or activity funded by the federal government.

APPROPRIATIONS BILL: Such bills are a special form of legislation that legally set aside money for a specific purpose. General appropriation bills originate in the House.

APPROPRIATIONS CYCLE: Once a separate congressional process, the yearly review and approval by Congress of spending levels for the following fiscal year is now integrated with the overall Congressional Budget Cycle. Appropriations committee hearings begin in February after the President submits his budget to the Congress and often continue through May and June. After the appropriations committees complete their hearings, mark-up (decisions on funding levels) begins. By July appropriation subcommittees report their bills to the floor of their respective houses and conference committee considerations begin. The current schedule calls for all actions on appropriation bills to be completed in time for presidential signature by September 30 of each year.

ASSIGNMENT PROCESS – BILLS: The decision by the Speaker of the House or the Majority Leader of the Senate determining which committee in their respective houses will handle a bill after its introduction. The assignment process often decides the fate of a bill.

ASSIGNMENT PROCESS – COMMITTEE: Appointment of a Member of Congress to various House or Senate Committees. Initial committee assignments are determined by special partisan committees in each house that recommend candidates to their majority or minority leaders.

---

* This glossary has been adapted from Pamela Ebert-Flattau's *A Legislative Guide* (prepared for distribution by the Association for the Advancement of Psychology, Washington, DC, 1980).

AUTHORIZATION: The Act of sanctioning federal support of a program or activity. In addition to specifying what a program is intended to do and who will do it, authorizing legislation often contains a section which provides for a fixed level of funding to meet program costs. While such funds may be authorized, no money may be spent unless it also has been appropriated from the Treasury through the separate legislative process of appropriating funds. Authorizing legislation is handled by the House and Senate through committees named for the subject matter over which they have jurisdiction, such as agriculture, armed services, foreign affairs, housing, nutrition, science, etc. Authorizing committees vary in their power and influence: some have tremendous power (e.g., Armed Services in both the House and Senate, Commerce Committees); others have less influence.

AUTHORIZATION BILL: The form of legislation used to incorporate the provisions of an authorization process for a program or activity.

AUTHORIZATION CYCLE: The review process involving approval/disapproval or creation/abolition by Congress of programs or activities proposed for the federal government to perform. Programs and activities may be authorized to operate for periods from one year to well over five years. At the time of its expiration, the program or activity is considered by Congress for reauthorization. Authorization hearings typically begin in February and continue through the months of March and April. Congressional authorizing committees must submit anticipated legislative authorizations to the Congressional Budget Office, and to the House or Senate Budget Committee, so that anticipated levels of outlays can be worked into the proposed Congressional budget. Authorizing committees of the House and Senate must report new authorizing legislation during May and June. Congress begins floor action on such legislation during July, which may continue through August.

BELTWAY: The circumferential highway around Washington, DC. Used to differentiate between things inside and outside of Washington.

BICAMERAL LEGISLATURE: Comprised of two co-equal houses. Congress is a bicameral legislature comprised of the House of Representatives and the Senate. Each house sets its own rules and proceedings.

BILL: One of the four principal forms used for legislation; besides the bill, there is the joint resolution, the concurrent resolution, and the simple resolution.

BRIEFING: A meeting in which a Member of Congress receives information pertinent to consideration of legislative or oversight action. A briefing may be conducted by a Member of the Senator's or Representative's staff, a committee staff member, a representative of one of the congressional support agencies, a lobbyist, an individual or group from a federal agency, a concerned constituent, or a representative of the scientific or engineering communities.

BRIEFING MEMO: A useful document for Members of Congress, generally a page in length. A typical briefing memo might include sections such as "Topic," "Background," "Key Issue," and "Recommended Action or Position To Take."

CALENDAR: The list on which all bills are placed as they are reported favorably by committees for House or Senate consideration. The calendar permits an orderly treatment of bills as they are presented to Congress. While there is only one calendar in the Senate, there are a number of calendars in the House of Representatives. Bills may remain on a calendar for the entire legislative session without being called up on the floor for debate, while others are there for only short periods of time.

A calendar of the House of Representatives, together with a history of all measures reported by a standing committee of either House, is printed each day the House is sitting. This information is made widely available for all who are interested. As soon as a bill (or other form of Congressional action) is favorably reported it is assigned a calendar number on either the Union Calendar or the House Calendar, the two principal calendars of business in the House. The calendar number is printed on the first page of the bill.

COMMITTEE CALENDAR: Each committee in the House and Senate periodically prepares an updated listing of the status of each piece of legislation that has been referred to the committee. This document is called the Committee Calendar.

CONSENT CALENDAR: If a measure pending before the House of Representatives is of a noncontroversial nature, it may be placed on the so-called Consent Calendar. On the first and third Monday of each month the Speaker of the House directs the clerk to call the bills that have been on the Consent Calendar for three days in the order of their appearance on that calendar. If objection is made to any bill, it is carried over on the calendar to the next day when the Consent Calendar is again called. If objection is made during this period by three or more Members it is stricken from the calendar and may not be placed on the Consent Calendar again in the same session of Congress. If no objection is heard if the bill is not "passed over" by request, it is passed by unanimous consent without debate. Ordinarily, the only amendments considered are those sponsored by the committee that reported the bill.

To avoid the passage without debate of measures that may be controversial or are sufficiently important or complex to require full discussion and debate there are six official objectors—three majority and three minority—who make a careful study of bills on the Consent Calendar. Objection to a bill that leads to its elimination from the Consent Calendar does not necessarily mean final defeat of the bill since it may then be brought up for consideration in the same way as any other bill on the House or Union Calendar in the House.

PRIVATE CALENDAR: Bills that affect an individual rather than the population at large. A private bill is used for legislative relief in matters such as immigration and naturalization and claims by or against the United States. A private bill is referred to the Private Calendar; this calendar is called up in the house on the first and third Tuesdays of each month.

UNION CALENDAR: Bills are referred to the calendar of the Committee of the Whole House on the State of the Union (commonly known as the Union Calendar) if they raise revenue, generally appropriate funds, or are of a public character and directly or indirectly appropriate money or property. The great majority of public bills and resolutions in the House are placed on this calendar.

WEDNESDAY CALENDAR: Another means for the House to dispense with legislation quickly. Each Wednesday, the Clerk calls the names of standing committees in alphabetical order. When called, a committee may raise for consideration any bill it reported on the previous day and pending on either the House or the Union calendar. Not more than two hours of general debate is permitted on the measure. The affirmative vote of a simple majority is sufficient to pass the bill.

CAUCUS: A group of persons, belonging to the same political party or faction, who meet to decide on policies and/or candidates.

CLOAKROOM: The anteroom to the House or the Senate chamber. In the old days, this literally referred to the place where Members would hang their cloaks.

CLOTURE: A vote of two-thirds of the Members of a house present needed to stop a filibuster and bring a bill to vote.

COMMITTEE: The House and Senate form committees to prepare legislation for action or to make investigations. Most standing committees are divided into subcommittees which study legislature, hold hearings, and report bills to the full committee, which can them report the legislation for action by the House and Senate.

CONCURRENT RESOLUTION: A matter affecting the operations of both houses of Congress is usually initiated in the form of a concurrent resolution. These are not customarily matters of legislative character but are merely expressions of fact, principle, opinion, or purpose of the two houses.

CONGRESSIONAL BUDGET AND IMPOUNDMENT CONTROL ACT OF 1974: Commonly called "the Congressional Budget Act," this Act revolutionized the Congressional budget process. The Congressional Budget Act was the result of a two-year study by Congress of the procedures that should be adopted for the purpose of improving Congressional control of budget outlay and receipt.

CONGRESSIONAL BUDGET CYCLE: Title III of the Congressional Budget Act establishes a timetable for various phases of the Congressional budget process and prescribes the actions to be taken at each point.

CONGRESSIONAL BUDGET OFFICE (CBO): A nonpartisan agency designed to provide Congress with information needed to make informed decisions about budget policy and national priorities. The CBO monitors the economy and estimates the impact on the economy of government actions; improves the flow and quality of budget information; and analyzes the costs and effects of alternative budget choices.

CONGRESSIONAL RECORD: The proceedings of the House and the Senate are printed and published daily in the *Congressional Record.*

CONGRESSIONAL RESEARCH SERVICE (CRS): Established in 1914 as the Legislative Reference Service, to provide Members of Congress, committees, and staff with information in an objective, nonpartisan and scholarly manner. Services include analysis of issues before Congress, legal research and analysis, consultation with Members of Congress, assistance with statements and speech drafts, and general reference assistance. CRS is a component of the Library of Congress.

CONSTITUENT: Every citizen, with the exception of those in the District of Columbia, is represented in Washington by one Member of the House of Representatives and two Senators to whom the constituent may turn for assistance or with ideas for legislation.

DISTRICT WORK PERIOD: A regularly scheduled period when the Senate or the House of Representatives is not in session.

DOCUMENT ROOM: The same night that a bill is introduced in either the House or the Senate, it is printed, and is available the next day from the document room of the appropriate house. Similarly, committee reports and public laws may be obtained, free of charge, from these document rooms. In order to get a document, send a self-addressed label and a note specifying the document by its appropriate prefix (e.g., H.R. 703) to:

Superintendent	Superintendent
Senate Document Room	House Document Room
SH-B04 Hart Senate	B-18 Ford House Office
Office Building	Building
Washington, DC 20510	Washington, DC 20515

Documents may be obtained in person by visiting these rooms located in the main Capitol Building.

ENGROSSED BILL: Once a bill has passed the House or the Senate, a copy is printed on blue paper with all the amendments in place. This engrossed bill is delivered in a rather formal ceremony to the other house while the body is actually sitting. Concurrence of the other house is thereby sought on the legislative matter.

EXPIRING LEGISLATION: In November of each year, Congressional staff examine the list of legislative items that expire in the following calendar year. This list is useful in gathering forces to lobby effectively for changes in the legislation. Expiring legislation is often the subject of committee and subcommittee hearings early in the next Congressional year.

FEDERAL REGISTER: A document published daily that includes federal agency regulations, proposed regulations, and other legal documents of the Executive Branch.

FILIBUSTER: A delaying tactic used by Senators to prevent a vote. Senator Strom Thurmond established the record for filibustering when he spoke against a civil rights bill in 1957 for 24 hours and 18 minutes. The technique may be used by a single dissident or a group of dissidents who have banded together to obstruct the passage of a bill. A filibuster may be stopped only by invoking cloture.

FISCAL YEAR: The period from October 1 to September 30 of government financial operations. Government-funded programs are authorized and funds appropriated coinciding with the fiscal year rather than the calendar year.

GALLERY: The balcony from which the public may observe the workings of the House or the Senate when they are in session.

GENERAL ACCOUNTING OFFICE (GAO): Headed by the Comptroller General of the United States, the General Accounting Office reviews and evaluates government programs carried on under existing law. This includes general review, evaluation, analysis and audit functions.

HEARING: If a bill is of sufficient importance, public hearings are scheduled. Announcements of hearings may be found in the Daily Digest portion of the *Congressional Record*. Personal notice, usually in the form of a letter, but possibly in the form of a subpoena, is sent to individuals, organizations and government departments and agencies that would be affected by the bill. Any interested party may ask to be heard. However, minor witnesses may be requested merely to submit written statements for the hearing record rather than to testify orally. Unless otherwise specified, hearings are open to the public. The product of a hearing, whether it is conducted by a committee or a subcommittee, is a transcript which is published as a separate, usually bound, document available for distribution.

HILL, THE: Congress and its associated agencies. Short for "Capitol Hill."

H.R.: The letters affixed to a bill introduced to the House of Representatives.

INTRODUCTION OF A BILL: In the House of Representatives, a Member simply drops a typed bill into what is called the hopper, thereby indicating its introduction. The Member is not required to ask permission to introduce the measure or to make any statement at the time of its introduction. In the Senate, the procedure is more formal. At the time reserved for such purposes, the Senator who wishes to introduce the measure rises and states that a bill is being offered for introduction. The bill is then sent by page to the Secretary's desk.

INTRODUCTORY REMARKS: The sponsoring Senator or Representative may ask permission to read or have printed in the Congressional Record remarks made at the time of a bill's introduction. These introductory remarks often contain statistics or other descriptive illustrations of need for the legislation. The scientist who has conducted a national survey, or who has information as a constituent pertinent to the remarks of the bill's sponsor, may find his or her facts or figures incorporated in the introductory remarks of the Member of Congress.

JOINT RESOLUTION: Joint resolutions may originate either in the House of Representatives or in the Senate, not jointly in both houses. There is little practical difference between a bill and a joint resolution although the latter is not as numerous as are bills. Joint resolutions may represent amendments to the Constitution, which must be approved by two-thirds of both houses and then sent to the States for their ratification.

LEGISLATIVE COUNSEL: A bill must state in precise legal language what it does and does not do. Hence, bill drafting has come to require a high degree of technical writing skill not possessed by the average citizen. Most Members of Congress call upon specialists to write legislation. After a Senator or Representative conceives the idea for a bill, the proposal is sketched out in rough draft and sent to the Legislative Counsel's office for transformation into legal language. Both the House and the Senate have their own legislative drafting staffs.

LOBBY: To exert influence on a Member of Congress to vote for or introduce legislation desired by an interest group. Also, an interest that engages in such activity.

MAJORITY LEADER: In the House of Representatives, the Majority Leader serves as the Speaker's chief lieutenant, handling partisan affairs. In the Senate, the Majority Leader is the chief Senate officer in practice, determining the Senate's legislative schedule and often leading the majority side in legislative debate.

MAJORITY WHIP: In the House the Majority Whip is third in command and is primarily responsible for polling the members of the party on pending legislation, informing them of bills and schedules, and summoning them to the floor for crucial votes. In the Senate, the Majority Whip fills in during the Majority Leader's absence from the Senate floor.

MARK-UP: Line-by-line, highly technical consideration of a bill by subcommittee or committee after hearings are completed. During the mark-up, views of both proponents and opponents of the legislation are studied in detail, legislative language is perfected, and a vote is taken of subcommittee or committee members to determine whether the bill should be reported favorably, reported unfavorably or tabled. Amendments to the legislation by subcommittee or committee members are also considered during the mark-up.

MINORITY LEADER: In the House, the Minority Leader is the highest ranking Member of the minority party; his or her duties include lining up opposition to legislation proposed by the majority party. In the Senate, the Minority Leader plays a similar role and often confers with the Majority Leader to obtain prior agreement on legislative scheduling, parliamentary tactics, and the like.

MINORITY WHIP: In both the House and the Senate, the Minority Whip assists the Minority Leader in the polling of Members on pending legislation, informing them of bills and schedules, and summoning them for crucial votes.

OPENING REMARKS: The statement made by the chairperson of a committee or a subcommittee at the time of hearings. The opening remarks lay out the goal of the hearings, as well as amplify on the content of the legislation ultimately considered by the Members of the committee or the subcommittee. As in the case of introductory remarks, opening remarks may utilize information provided by experts to document the need for a public hearing.

OVERSIGHT: The act of reviewing and monitoring federal programs and policies. Oversight responsibilities typically belong to standing committees in various jurisdictional areas in the House and in the Senate.

POCKET VETO: The President has ten days in which to sign a bill that has been passed by both the House and the Senate. If he signs it, it becomes law. If he does not approve it, he can veto it and send it back to the originating house. If the President does not wish to sign the bill or to veto it, and the Congress adjourns before the ten-day period elapses, the bill fails to become law through a "pocket veto."

P.L.: A bill that becomes a public law is assigned the letters P.L. and given a number. The first two digits indicate the number of the Congress in which the law was enacted.

PRESIDENT OF THE SENATE: The Vice President of the United States presides over the Senate but is not a Member of any standing committees and has minimal powers. The President of the Senate usually presides only during important debates.

PRESIDENT PRO TEMPORE: The senior Member of the majority party in the Senate is usually elected as President Pro Tempore. The President Pro Tempore presides over the Senate in the Vice-President's absence.

RESCISSION: Literally, a taking away. The act of canceling or voiding, usually of appropriations of government funds.

REPORT: An official document of a Senate or House committee or subcommittee, that summarizes the findings and actions taken by the committee or subcommittee pursuant to hearings, mark-up, and committee discussion of a bill. Reports referred to the House or to the Senate for consideration are given numbers much like numbers given to bills and may be obtained through House or Senate document offices in the manner described elsewhere.

REPRESENTATIVE: Individuals elected to membership in the U.S. Congress for a period of two years, representing a district of approximately 500,000 citizens. A Representative must be at least 25 years of age, have been a citizen of the U.S. for seven years and, when elected, have been a resident of the State in which he or she is chosen.

ROLL CALL: Casting votes in the Senate or in the House of Representatives by calling the name of each Member. Roll call votes are published in the Congressional Record on the day after the vote.

RULE: An ordinary bill reported out by Committee and assigned to a calendar in the House of Representatives must still clear the Committee on Rules before it reaches the House floor. The Committee on Rules was established to determine which bills deserve to proceed and what ground rules will be used for debate. A bill that has been cleared by the Rules Committee is accompanied by a resolution that specifies the rules for debating the bill. The resolution itself is debated for up to one hour before the bill is brought up. Rules vary but some typical ones include a rule that says a bill will be debated for two hours after which the bill is open to amendments, or a rule that says a controversial bill may have from 8 to 10 hours for general debate.

S.: A bill introduced in the Senate is designated by the letter "S." followed by a number.

SENATOR: One of two individuals elected from each state of the union to serve in the U.S. Congress for a period of six years. A Senator must be at least 30 years old, have been a citizen of the United States for nine years, and when elected, must be a resident of the State from which he or she is chosen.

SESSION: A calendar year. Congress convenes for two years, the first year of which is referred to as the "first session" and the second year, the "second session."

SPEAKER: The chief of the House of Representatives. The Speaker is chosen by the majority party in a presession caucus and is formally nominated along with the minority party candidate, as the first order of business in the first session of a new Congress. The Speaker is the leader of the majority party, but is expected to remain impartial in rulings while presiding over the House. The Speaker has broad powers over legislative scheduling, committee appointments, rules during debate, and recognition of Members from the floor.

SPONSORSHIP: A bill introduced in the House of Representatives may have no more than 25 Members signing as cosponsor. In the Senate, sponsorship is unlimited. Occasionally, a Member may insert the words "by request" after his or her name to indicate that cosponsorship is in compliance with the suggestion of some other Member, or the Administration, and does not necessarily reflect the commitment of the sponsor.

STAFF: Staff frame the questions for their bosses' consideration, schedule committee hearings, select witnesses, meet with lobbyists and interested constituents to discuss legislation, and respond to enormous volumes of constituent mail.

STATE DELEGATION: The total number of Representatives and Senators elected from a State.

STATUTES AT LARGE: Existing legislation.

TABLE: A parliamentary motion to remove a bill from consideration. A bill or legislative proposal may be tabled by a subcommittee or committee for the remainder of a Congressional session.

TESTIFY: To make a formal statement in a committee or subcommittee hearing in favor of or opposition to a legislative measure. Individuals invited to testify before a committee or subcommittee are requested to appear as witnesses in person having filed their written statement in advance. After the presentation of their testimony, witnesses are cross-examined by members of the committee or subcommittee holding the hearing.

VETO: Once a bill has been approved by both houses, it becomes "an enrolled bill" and is sent to the White House for the signature of the President. The President has ten days in which to sign a bill. If he does not approve it, he can veto it and send it back to the originating house. A two-thirds majority of each house is needed to override a veto, at which time it become law despite the President's veto. (See also Pocket Veto.)

WHIP: See Majority Whip, Minority Whip.

APPENDIX A

ADDRESSES AND TELEPHONE NUMBERS FOR KEY
CONGRESSIONAL COMMITTEES AND OTHER
OFFICES

I. Senate Committees and Subcommittees
   Concerned with Science and Technology

Address a letter to a Senator with:
The Honorable (full name)
U.S. Senate
Washington, DC 20510

Dear Senator (last name):

Senate Internet addresses can be found at
http://www.senate.gov or gopher.senate.gov.

Committee on Agriculture, Nutrition, and Forestry
SR-328A Russell Senate Office Building
U.S. Senate
Washington, DC 20510
202/224-2035, Fax: 202/224-1725
Use the full committee address and phone number
for the subcommittees.

   **Subcommittee on Forestry, Conservation,
   and Rural Revitalization**

   **Subcommittee on Research, Nutrition, and
   General Legislation**

Committee on Appropriations
S-128 Capitol
U.S. Senate
Washington, DC 20510
202/224-3471, Fax: 202/224-4168

Subcommittee on Agriculture, Rural
Development, and Related Agencies
SD-136 Dirksen Senate Office Building
202/224-5270

Subcommittee on Commerce, Justice, State,
the Judiciary, and Related Agencies
S-146A Capitol
202/224-7277

Subcommittee on Defense
SD-122 Dirksen Senate Office Building
202/224-7255

Subcommittee on Energy and Water
Development
SD-131 Dirksen Senate Office Building
202/224-7260

Subcommittee on Interior and Related
Agencies
SD-127 Dirksen Senate Office Building
202/224-7233

Subcommittee on Labor, Health and Human
Services, Education, and Related Agencies
SD-184 Dirksen Senate Office Building
202/224-7230

Subcommittee on Transportation and
Related Agencies
SD-133 Dirksen Senate Office Building
202/224-7281

Subcommittee on Veterans' Affairs, Housing
and Urban Development, and Independent
Agencies
SD-131 Dirksen Senate Office Building
202/224-2833

Committee on Armed Services
SR-228 Russell Senate Office Building
U.S. Senate
Washington, DC 20510
202/224-3871, Fax: 202/228-3781

Use full committee address and phone number for the subcommittee.

Subcommittee on Acquisition and Technology

Subcommittee on Strategic Forces

Committee on the Budget
SD-621 Dirksen Senate Office Building
U.S. Senate
Washington, DC 20510
202/224-0642, Fax: 202/224-4835

Committee on Commerce, Science, and Transportation
SR-254 Russell Senate Office Building
U.S. Senate
Washington, DC 20510
202/224-1251, Fax: 202/228-1259

Subcommittee on Aviation
SH-427 Hart Senate Office Building
202/224-4852, Fax: 202/228-0326

Subcommittee on Communications
SH-227 Hart Senate Office Building
202/224-5184, Fax: 202/224-9334

Subcommittee on Oceans and Fisheries
SH-428 Hart Senate Office Building
202/224-8172, Fax: 202/228-0326

Subcommittee on Science, Technology, and Space
SH-428 Hart Senate Office Building
202/224-8172, Fax: 202/228-0326

Subcommittee on Surface Transportation and Merchant Marine
SH-427 Hart Senate Office Building
202/224-4852, Fax: 202/224-0326

Committee on Energy and Natural Resources
SD-364 Dirksen Senate Office Building
U.S. Senate
Washington, DC 20510
202/224-4971, Fax: 202/224-6163

Subcommittee on Energy Production and Regulation
SD-308 Dirksen Senate Office Building
202/224-6567, Fax: 202/228-0302

Subcommittee on Energy Research and Development
SD-308 Dirksen Senate Office Building
202/224-8115, Fax: 202/228-0302

Subcommittee on Forests and Public Land Management
SD-306 Dirksen Senate Office Building
202/224-6170, Fax: 202/228-0539

Committee on Environment and Public Works
SD-410 Dirksen Senate Office Building
U.S. Senate
Washington, DC 20510
202/224-6176, Fax: 202/224-5167
Use the full committee address and phone number for the subcommittees.

Subcommittee on Clean Air, Wetlands, Private Property, and Nuclear Safety

Subcommittee on Drinking Water, Fisheries, and Wildlife

Subcommittee on Superfund, Waste Control, and Risk Assessment

Subcommittee on Transportation and Infrastructure

Committee on Foreign Relations
SD-450 Dirksen Senate Office Building
U.S. Senate
Washington, DC 20510
202/224-4651

Use the full committee address and phone number for the subcommittees.

Subcommittee on International Economic Policy, Export, and Trade Promotion

Committee on Labor and Human Resources
SD-428 Dirksen Senate Office Building
U.S. Senate
Washington, DC 20510
202/224-5375, Fax: 202/224-5044

II. House of Representatives Committees and Subcommittees Concerned with Science and Technology

Address a letter to a Representative with:
The Honorable (full name)
U.S. House of Representatives
Washington, DC 20515

Dear Congressman or Congresswoman (last name):

Internet addresses for House committees and subcommittees can be located at http://www.house.gov or gopher.house.gov.

Committee on Agriculture
1301 Longworth House Office Building
U.S. House of Representatives
Washington, DC 20515
202/225-2171, Fax: 202/225-0917
Use the full committee address and phone number for the subcommittees.

Subcommittee on Department Operations, Nutrition, and Foreign Agriculture

Subcommittee on Resource Conservation, Research, and Forestry

Committee on Appropriations
H-218 Capitol
U.S. House of Representatives
Washington, DC 20515
202/225-2771, Fax: 202/225-5078

Subcommittee on Agriculture, Rural
Development, the Food and Drug
Administration, and Related Agencies
2362 Rayburn House Office Building
202/225-2638

Subcommittee on Commerce, Justice, State,
the Judiciary, and Related Agencies
H-309 Capitol
202/225-3351

Subcommittee on Energy and Water
Development
2362 Rayburn House Office Building
202/225-3421

Subcommittee on Interior and Related
Agencies
B-308 Rayburn House Office Building
202/225-3081

Subcommittee on Labor, Health and Human
Services, Education, and Related Agencies
2358 Rayburn House Office Building
202/225-3508

Subcommittee on National Security
H-149 Capitol
202/225-2847

Subcommittee on Transportation and
Related Agencies
2358 Rayburn House Office Building
202/225-2141

Subcommittee on Veterans' Affairs, Housing
and Urban Development, and Independent
Agencies
H-143 Capitol
202/225-3241

Committee on Budget
309 Cannon House Office Building
U.S. House of Representatives
Washington, DC 20515
202/226-7270, Fax: 202/226-7174

Committee on Commerce
2125 Rayburn House Office Building
U.S. House of Representatives
Washington, DC 20515
202/225-2927, Fax: 202/225-1919
Use the full committee address and phone numbers
for the subcommittees.

Subcommittee on Commerce, Trade, and Hazardous Materials

Subcommittee on Energy and Power

Subcommittee on Health and Environment

Subcommittee on Oversight and Investigations

Subcommittee on Telecommunications and Finance

Committee on Economic and Educational Opportunities
2181 Rayburn House Office Building
U.S. House of Representatives
Washington, DC 20515
202/225-4527, Fax: 202/225-9571

Committee on International Relations
2170 Rayburn House Office Building
U.S. House of Representatives
Washington, DC 20515
202/225-5021

Committee on National Security
2120 Rayburn House Office Building
U.S. House of Representatives
Washington, DC 20515
202/225-4151, Fax: 202/225-9077

Subcommittee on Military Research and Development
2340 Rayburn House Office Building
202/225-0883, Fax: 202/226-0105

Committee on Resources
1324 Longworth House Office Building
U.S. House of Representatives
Washington, DC 20515
202/225-2761, Fax: 202/225-5929

Subcommittee on Fisheries, Wildlife, and Oceans
805 O'Neill House Office Building
202/226-0200, Fax: 202/225-1542

Subcommittee on Energy and Mineral Resources
1626 Longworth House Office Building
202/225-9297, Fax: 202/225-5255

Subcommittee on Water and Power Resources
1337 Longworth House Office Building
202/225-8331, Fax: 202/225-6953

Committee on Science
2320 Rayburn House Office Building
U.S. House of Representatives
Washington, DC 20515
202/225-6371, Fax: 202/226-0891

Subcommittee on Basic Research
2319 Rayburn House Office Building
202/225-9662, Fax: 202/225-7815

Subcommittee on Energy and Environment
B-374 Rayburn House Office Building
202/225-9662, Fax: 202/226-6983

Subcommittee on Space and Aeronautics
2320 Rayburn House Office Building
202/225-7858, Fax: 202/225-6415

Subcommittee on Technology
B-374 Rayburn House Office Building
202/225-8844, Fax: 202/225-4438

Committee on Transportation and Infrastructure
2165 Rayburn House Office Building
U.S. House of Representatives
Washington, DC 20515
202/225-9446 , Fax: 202/225-6782

> **Subcommittee on Aviation**
> 2251 Rayburn House Office Building
> 202/226-3220, Fax: 202/225-4629

> **Subcommittee on Surface Transportation**
> B-370A Rayburn House Office Building
> 202/225-6715, Fax: 202/225-4623

> **Subcommittee on Water Resources and Environment**
> B-376 Rayburn House Office Building
> 202/225-4360, Fax: 202/226-5435

III. Congressional Support Agencies Concerned with Science and Technology

Congressional Budget Office
Ford House Office Building, Fourth Floor
Washington, DC 20515
202/226-2621

Congressional Research Service
Library of Congress
James Madison Memorial Building
101 Independence Avenue and South Capitol Street, SW
Washington, DC 20515
202/707-5700

> **Environment and Natural Resources Policy Division**
> 202/707-7232

> **Science Policy Research Division**
> 202/707-9547

General Accounting Office
441 G Street, NW
Washington, DC 20548
202/512-4800
(To order GAO reports call 202/512-6000.)

> **Resources, Community, and Economic Development Division**
> 202/512-3200

IV. Congressional Leadership Organizations

Senate Leadership Organizations
Senate Democratic Conference
S-309 Capitol Building
202/224-3735

> **Senate Democratic Policy Committee**
> S-118 Capitol Building
> 202/224-5551

> **Senate Democratic Steering and Coordination Committee**
> SH-619 Hart Senate Office Building
> 202/224-3232

> **Democratic Technology and Communications Committee**
> 712 Senate Hart Office Building
> 202/224-1430

> **Senate Republican Conference**
> SH-405 Hart Senate Office Building
> 202/224-2764

> **Senate Republican Policy Committee**
> SR-347 Russell Senate Office Building
> 202/224-2946, Fax 202/224-1235

House of Representatives Leadership Organizations
House Democratic Caucus
1420 Longworth House Office Building
202/226-3210, Fax: 202/225-0282

House Democratic Policy Committee
H-301 Capitol Building
202/225-6760

House Republican Conference
1010 Longworth House Office Building
202/225-5107, Fax: 202/225-0809

House Republican Policy Committee
21471 Rayburn House Office Building
202/225-6168, Fax: 202/225-0931

Informal House Organizations
Business for informal organizations of House members is often handled through the offices of the group's co-chairs. Call the House switchboard to determine whom to contact, 202/225-3121.

Congressional Aviation and Space Caucus
Congressional Biomedical Research Caucus
Forestry 2000 Task Force
Medical Technology Caucus

Bicameral Organizations
Business for congressional bicameral organizations is often handled through the offices of the group's co-chairs. Call the House or Senate switchboards, 202/225-3121 or 202/224-3121 respectively, to determine whom to contact.

Congressional Biotechnology Caucus

Congressional Caucus on Advanced Materials

Congressional Competitiveness Caucus

APPENDIX B

SUGGESTED READINGS

I. The Governmental Environment:
    A Focus on the Congress

Bacon, Donald C., Roger H. Davidson, and Morton Keller, eds. *The Encyclopedia of the United States Congress.* New York: Academic PR., Inc., 1994.

Burns, James M., J.W. Peltason, and Thomas E. Cronin. *Government by the People.* 6th ed. Englewood Cliffs: Prentice-Hall, 1989.

Carnegie Commission on Science, Technology, and Government. *Science, Technology, and Congress: Analysis and Advice from the Congressional Support Agencies.* New York: Carnegie Commission on Science, Technology, and Government, 1991.

Carnegie Commission on Science, Technology, and Government. *Science, Technology, and Congress: Expert Advice and the Decision-Making Process.* New York: Carnegie Commission on Science, Technology, and Government, 1991.

Chubb, John E., and Paul E. Peterson, eds. *Can the Government Govern?* Washington, DC: Brookings Institution, 1989.

Congressional Quarterly. *How Congress Works.* Washington, DC: Congressional Quarterly, 1991.

Davidson, Roger H., and Walter J. Oleszek. *Congress and Its Members.* 4th ed. Washington, DC: Congressional Quarterly, 1993.

Dodd, Lawrence, and Calvin Jillson. *New Perspectives on American Politics.* Washington, DC: Congressional Quarterly, 1994.

Dodd, Lawrence C., and Bruce I. Oppenheimer, eds. *Congress Reconsidered.* 5th ed. Washington, DC: Congressional Quarterly, 1993.

Duncan, Phil, ed. *Politics in America: 1996.* Washington, DC: Congressional Quarterly, 1995.

Dye, Thomas R., and L. Harmon Zeigler. *The Irony of Democracy: An Uncommon Introduction to American Politics.* 10th ed. Monterey: International Thomson Publishing, 1996.

Elving, Ronald D. *Conflict and Compromise: How Congress Makes the Law.* New York: Simon & Schuster, 1995.

Fisher, Louis. *The Politics of Shared Power: Congress and the Executive.* 3rd ed. Washington, DC: Congressional Quarterly, 1992.

Gore, Al. *Creating a Government that Works More and Costs Less: The Report of the National Performance.* NAL-Dutton, 1993.

Graduate Group. *First Annual What Your Congressman Can Do For You.* West Hartford: Graduate Group, 1996.

Harris, Fred R. *In Defense of Congress.* New York: St. Martin's Press, 1994.

Heinz, John P., et al. *The Hollow Core: Private Interests in National Policy Making.* Cambridge: Harvard University Press, 1993.

Hess, Carol, ed. *Political Resource Directory: National Edition 1994.* 7th ed. Rye, NY: Political Resource Directories, 1994.

Jones, Bryan D., ed. *The New American Politics? Reflections on Political Change, the 1992 Election, and the Clinton Administration.* Boulder: Westview Press, 1995.

Mann, Thomas E. and Norman J. Ornstein. *Congress, the Press, and the Public.* Washington, DC: American Enterprise Institute and the Brookings Institution, 1994.

Mann, Thomas E. and Norman J. Ornstein, directors. *A Second Report of the Renewing Congress Project.* Washington, DC: American Institute for Public Policy Research and the Brookings Institution, 1993.

Maxwell, Bruce. *How to Access the Federal Government on the Internet 1995: Washington Online.* Washington, DC: Congressional Quarterly, December 1995.

Oleszek, Walter J. *Congressional Procedures: The Policy Process.* ed. Washington, DC: Congressional Quarterly, 1996.

Rieselbach, L. *Congressional Politics: The Evolving Legislative System.* 2nd ed. Boulder, CO: Westview Press, 1995.

Rieselbach, L. *Congressional Reform: The Changing Modern Congress.* Washington, DC: Congressional Quarterly, 1993.

Smith, Steven S. *Committees in Congress.* 2nd ed. Washington, DC: Congressional Quarterly, 1990.

Sundquist, James L. *Constitutional Reform and Effective Government.* Washington, DC: Brookings Institution, 1986.

Wittenberg, Ernest, and Elisabeth Wittenberg. *How to Win in Washington: Very Practical Advice about Lobbying, the Grassroots, and the Media.* Cambridge: Blackwell Pubs., 1990.

Woods, Patricia D. *The Dynamics of Congress: A Guide to the People and Process in the U.S. Congress.* Washington, DC: Woods Institute, 1991.

II. Preparing and Giving Testimony and Presentations

Anholt, Robert R.H. *Dazzle 'Em With Style: An Introduction to the Art of Oral Scientific Presentation.* New York: W. H. Freeman and Company, 1995.

Arrendo, Lani. *How to Present Like a Pro: Getting People to See Things Your Way.* New York: McGraw-Hill, 1991.

Bedrosian, Margaret McAuliffe. *Speak Like a Pro: In Business and Public Speaking.* New York: Wiley, 1987.

Bell, Arthur H.; Skopek, Eric W. *The Speaker's Edge: Tips for Confident Presenting.* Westbury: Asher-Gallant Press, 1988.

Boylan, Bob. What's Your Point? *A Proven Method for Giving Crystal Clear Presentations!* Wayzata: Point Publications, 1989.

Brooks, William T. *High Impact Public Speaking.* Englewood Cliffs: Prentice Hall, 1988.

Echeverria, Ellen W. *Speaking on Issues: An Introduction to Public Communication.* Orlando: Harcourt Brace College Pubs., 1987.

Gard, Grant C. *The Art of Confident Public Speaking.* Englewood Cliffs: Prentice Hall, 1986.

Gilbert, Frederick. *PowerSpeaking: How Ordinary People Can Make Extraordinary Presentations.* Redwood City: Frederick Gilbert Associates, 1994.

Harvey, Thomas A. "Congressional Testimony: A Practical Guide for Newcomers to the Congressional Hearing Process." *Federal Bar News & Journal* 35 (Nov. 1988): 414-15.

Johnson, Carl, et. al. *Choices: Decision Making Processes for Speakers.* 2nd ed. Kendall-Hunt, 1993.

Kenny, Peter. *A Handbook of Public Speaking for Scientists and Engineers.* Philadelphia: IOP Publishing Co., 1982.

Kougl, Kathleen M. *Primer for Public Speaking.* New York: HarperCollins College, 1990.

Leeds, Dorothy. *PowerSpeak.* New York: Berkley Publishing Group, 1991.

Lustberg, Arch. *Testifying With Impact.* Washington, DC: Association Division, U.S. Chamber of Commerce, 1982.

Minninger, Joan, and Barbara Goulter. *The Perfect Presentation.* New York: Doubleday, 1991.

Osgood, Charles. *Osgood on Speaking: How to Think on Your Feet Without Falling on Your Face.* New York: Morrow, 1988.

Slim, Hugo, et al. *Listening for a Change: Oral Testimony and Development.* Philadelphia: New Society Pubs., 1994.

Smith, Terry C. *Making Successful Presentations: A Self-Teaching Guide.* 2nd ed. New York: Wiley, 1991.

Snyder, Marilyn. *High Performance Speaking.* Burr Ridge: Irwin Professional Publishing, 1994.

Vasile, Albert J.; Mintz, Harold K. *Speak With Confidence: A Practical Guide.* 5th ed. Glenview: HaperCollins College, 1989.

APPENDIX C

SOURCES OF INFORMATION ABOUT CONGRESS

I. Reference Books

Almanac of American Politics. Profiles of senators, representatives, and governors, with voting records, election results, and district information; published annually by the National Journal, 1501 M St. NW, Washington, DC 20005; 202/739-8400; $64.95 hardcover, $49.95 softcover, plus 10 percent shipping; available in some bookstores.

Congressional Staff Directory. Comprehensive reference to key congressional staffers, including biographies; published bi-annually by Staff Directories, Ltd., Mount Vernon, VA 22121-0062; 703/739-0900, Fax: 703/739-0234; $79; also available from Trover Shop Books, in Washington, DC.

Congressional Yellow Book. Comprehensive reference guide of Members of Congress and committees; published quarterly by Leadership Directories, Inc., 1301 Pennsylvania Avenue, NW, Suite 925, Washington, DC 20004; 202/347-7757, Fax: 202/628-3430; annual subscription rate is $235.

Politics in America. State-by-state guide, with both national and state profiles and statistics; published annually by Congressional Quarterly Press, 1414 22nd Street, NW, Washington, DC 20037; 202/887-8500; $89.95 hardcover; $54.95 softcover; available in some bookstores.

Senate Telephone Directory. Listings of senators and personal and committee staff; published semi-annually; available from the Superintendent of Documents at the U.S. Government Printing Office, Washington, DC 20402; 202/512-1800; $12.

U.S. House of Representatives Telephone Directory. Listings of U.S. Representatives and personal and committee staff; updated two or three times a year; available from the Superintendent of Documents at the U.S. Government Printing Office, Washington, DC 20402; 202/512-1800; $15.

II. Periodicals and Newsletters

Aviation Week & Space Technology. Weekly news magazine covering national security, arms control, space, and commercial aviation issues; available from McGraw Hill, Inc., 1221 Avenue of the Americas, New York, NY 10020; 212/512-2000 or 1-800-525-5003; $82 per year.

Chemical & Engineering News. Weekly magazine; published by the American Chemical Society, 1155 16th Street, NW, Washington, DC 20036; 202/872-4600; contact Member & Subscriber Services, ACS, PO Box 3337, Columbus, OH, 43210; 800/333-9511; $120 per year.

Congressional Monitor. Daily newsletter providing an overview of the day's expected actions, updates of the previous day's action, a schedule of committee meetings, and status of bills; published by Congressional Quarterly, Inc., 1414 22nd Street, NW, Washington, DC 20037; 202/887-6279; $1,399 per year.

Congressional Record. Daily report of the public proceedings of the House of Representatives and the Senate, published each day that one or both chambers are in session; printed by the U.S. Government Printing Office, Superintendent of Documents, Washington, DC 20402; $225 per year. Also available online through GPO Access free of charge (http://www.access.gpo.gov/su_docs); for general information about GPO Access, send an e-mail message to help@eids05.eids.gpo.gov or by calling 202/512-1530.

Congressional Quarterly Weekly Report. Weekly publication covering major issues coming before Congress (including votes and status of legislation); published by Congressional Quarterly, Inc., 1414 22nd Street, NW, Washington, DC 20037; 202/887-8500 or 1-800-638-1710; $1,395 per year.

Environment and Energy Weekly Bulletin. Weekly publication covering congressional action on environment, energy, and natural resources legislation; available from Congressional Green Sheets, Inc., 406 E Street, SE, Washington, DC 20003; 202/546-2220, Fax: 202/546-7490, e-mail: wb@cais.com; $695 per year.

Environmental and Energy Study Institute Weekly Bulletin. Weekly publication providing background, history, and overview of current debate on environmental and energy issues before Congress; available from the Environmental and Energy Study Institute, 122 C Street, NW, Suite 700, Washington, DC 20001; 202/628-1400; $345 per year.

FYI. Periodic "electronic update" on issues of interest to the S&T community (may also be available in print form); available from the American Institute of Physics, One Physics Ellipse, College Park, MD 20740; 301/209-3108, Fax: 301/209-0846.

Issues in Science & Technology. Quarterly journal sponsored by the National Academy of Sciences, the National Academy of Engineering, the Institute of Medicine, the Cecil and Ida Green Center for the Study of Science and Society, and the University of Texas at Dallas; contact Customer Service, Issues in Science and Technology, PO Box 661, Holmes, PA 19043; 214/883-6325; issuessn@utdallas.edu; $43.50 per year.

National Journal. Weekly publication with in-depth articles and analysis of politics and government; available from National Journal,
1501 M Street, NW, Washington, DC 20005; 202/739-8400; $889 per year.

Nature. Weekly British scientific journal; published by Macmillan Magazines, Ltd.; 202/737-2355; contact Nature, Subscription Department, PO Box 1733, Riverton, NJ 08077-7333; $145 per year.

Roll Call. Capitol Hill's local semiweekly newspaper; published by Roll Call, Inc., 900 Second Street, NE, Suite 107, Washington, DC 20002; 202/289-4900; $210 per year.

Science. Weekly journal of the American Association for the Advancement of Science; contact AAAS Membership and Circulation, 1200 New York Avenue, NW, Washington, DC 20005; 202/326-6400; subscription is included in $102 annual dues for members.

Science & Government Report. Bi-monthly newsletter of science policy issues; published by Science & Government Report, Inc., 3736 Kanawha Street, NW, Washington, DC 20015; 202/785-5054; $455 per year.

Science & Technology in Congress. Monthly newsletter covering current congressional action on S&T issues; published by the AAAS Center for Science, Technology, and Congress, 1200 New York Avenue, NW, Washington, DC 20005; 202/326-6600 or congress_center@aaas.org.

The Scientist. Biweekly life sciences newspaper published by The Scientist, Inc.; contact The Scientist, PO Box 10525, Riverton, NJ 08076; $29 per year.

Washington Technology. Biweekly newspaper published by Tech News, Inc., 1953 Gallows Road, Vienna, VA 22180; $49 per year.

What's New. Periodic "electronic update" on items of interest to the S&T community (may also be available in print form); available from the American Physical Society, 529 14th Street, NW, Suite 1050, Washington, DC 20045; 202/662-8700, Fax: 202/667-8711, opa@aps.org.

In addition, many of the professional societies listed in Appendix F publish discipline specific newsletters that contain policy-oriented information. Contact the individual societies for information about their newsletters.

III. Internet Sources

The number of World Wide Web sites that deal with the intersection between science, technology, and Congress (or government) grows almost daily. Here are a few that provide information about congressional activities and useful links to other sites.

Thomas (http://thomas.loc.gov/) is a legislative information service of the Library of Congress providing the full-text of legislation introduced in the 103rd and 104th Congress and bill summary and status as well as links to C-Span and other Library of Congress Government Resources.

The U.S. House of Representatives' World Wide Web site (http://www.house.gov/) provides public access to legislative information as well as information about member of the House, House committees, and organizations of the House. It also provides links to other U.S. government information resources.

The U.S. Senate's World Wide Web site (http://www.senate.gov/) provides information from and about the members of the Senate, Senate committees, and Senate leadership and support offices. It also provides general background information about U.S. Senate legislative procedures, Senate facilities in the Capitol Building, and the history of the Senate.

The White Office of Science and Technology Policy's World Wide Web site (http://www.whitehouse.gov/White_House/EOP/OSTP/html/OSTP_home.html) provides information about activities at OSTP, OSTP documents and testimony, and White House publications dealing with science and technology. It also provides links to other resources.

The American Association for the Advancement of Science's World Wide Web site (http://www.aaas.org/) provides information about congressional appropriations for research and development and information from the AAAS Center for Science, Technology, and Congress.

PoliticsUSA (http://www.politicsusa.com/), produced by National Journal, Inc. and the American Political Network, provides information about political news, resources, and involvement. The information includes current information on congressional and Administration activity.

Science and Technology Policy World Wide Web site (http://www.cchem.berkeley.edu/~pabgrp/People/bob/policy.html), produced by Ph.D. candidate Bob Rich at the University of Berkeley's Department of Chemistry, provides access to a variety of science and technology policy resources throughout the Web.

Yahoo (http://www.yahoo.com/) is a subject-oriented database guide for the World Wide Web and Internet. It includes a subject area for science and provides links to other web sites.

APPENDIX D
OBTAINING CONGRESSIONAL DOCUMENTS

These contacts should be useful in acquiring congressional documents. When calling or visiting, have specific information about the desired document, such as document's title, author(s), or publication number, ready.

Senate Document Room
Superintendent of Documents
SH-B04 Hart Senate Office Building
Washington, DC 20510
202/224-7860

House Document Room
Superintendent of Documents
B-18 Ford House Office Building
Washington, DC 20515
202/225-3456
Internet: hdocs@hr.house.gov

When ordering, have report titles and public law or bill numbers ready. You may order up to six different documents from the House and Senate Document Rooms at no charge. Multiple copies of documents are available from the Government Printing Office for a fee.

U.S. Government Printing Office (GPO)
Superintendent of Documents
202/512-1800
In person: 710 North Capitol Street, NW
Washington, DC 20401
202/512-0132

or

By mail: 1510 H Street, NW
Washington, DC 20005
202/653-5075

Call before requesting by mail to be sure that the document is in stock. Orders should include title of publication and GPO stock number if possible. A check or money order made out to the Superintendent of Documents should accompany orders or orders may be charged to Visa or Mastercard. GPO also handles subscriptions to the Congressional Record and the Federal Register, and maintains a catalog of U.S. Government Publications. Free copies of U.S. Government Books, New Books List, and Government Periodicals and Subscription Services are available.

General Accounting Office (GAO)
Document Handling and Information Services Directory
700 4th Street, NW
Washington, DC 20548
202/512-6000

The GAO publishes a monthly list of its reports. The first copy of a report is available at no cost; there is a charge for additional copies.

Congressional Budget Office (CBO)
Ford House Office Building, 4th Floor
Washington, DC 20515
202/226-2809

A free publication list is available from CBO. The first copy of a report is available at no cost; there is a charge for additional copies.

Congressional Research Service (CRS)
Library of Congress
James Madison Memorial Building
101 Independence Avenue and South Capitol Street, SW
Washington, DC 20515
202/707-5700

CRS reports are available only through congressional offices.

Appendix E

Library of Congress

As the nation's main repository, the Library of Congress is an invaluable source of information. The Library maintains collections of journals, newspapers, manuscripts, films, photos, maps, presidential papers, and recorded speeches as well as books. Researchers must show photo identification to request materials in all of the Library's public reading rooms.

Library of Congress
10 First Street, SE
Washington, DC 20540
202/707-5000

Reference Services: 202/707-5522
Reference services are available by phone, by mail, or in person. There are various restrictions and time constraints (e.g. , some materials are off-site and require several days to acquire).

Loan Division: 202/707-5441
Generally, Library of Congress materials are for "in library" use; however, some materials may be obtained through interlibrary loans.

Photo Duplication Services: 202/707-5640
There is a fee for photo duplication services, and the services are subject to restrictions.

Reading Rooms
Call 202/707-6400 for recorded information on reading room hours of operations.

Government Publications Reading Room
Madison Building, Room 133
202/707-5647

Library Reading Room
Madison Building, Room 201
202/707-5099

Main Reading Room
Jefferson Building, Room 100
202/707-5522

Newspaper and Current Periodical Reading Room
Madison Building, Room 133
202/707-5690

Science Reading Room
Adams Building, Room 5010A
202/707-5639

APPENDIX F

Congressional Quarterly's *Washington Information Directory* (published annually) is a good source of information, including addresses and telephone numbers for many of these organizations.

I.   Professional Societies

American Academy of Physician Assistants
950 N. Washington Street, Alexandria, VA 22314
703/836-2272, Fax: 703/684-1924
Contact: Stephen C. Crane, Executive
Vice-President

American Anthropological Association
4350 N. FairFax Drive, Suite 640,
Arlington, VA 22203
703/528-1902, Fax: 703/528-3546
Contact: Jack Cornman, Executive Director

American Association for the Advancement of Science
1200 New York Avenue, NW
Washington, DC 20005
202/326-6600, Fax: 202/289-4950
Contact: Albert H. Teich, Director, Science and
Policy Programs

American Association for Clinical Chemistry
2101 L Street NW, Suite 202
Washington, DC 20037
202/857-0717; 800/892-1400, Fax: 202/887-5093
Contact: Richard Flaherty, Executive
Vice-President

American Association of Engineering Societies
415 2nd Street, NE, Suite 200
Washington, DC 20002
202/296-2237, Fax: 202/296-1151
Contact: Brian Dougherty, Manager, Government
Relations

American Astronomical Society
2000 Florida Avenue, NW, Suite 300
Washington, DC 20036
202/328-2010, Fax: 202/234-2560
Contact: Peter B. Boyce, Senior Associate

American Chemical Society
1155 16th Street, NW, Washington, DC 20036
202/872-4477, Fax: 202/872-6206
Contact: Susan M. Turner, Head, Government
Relations and Science Policy

American Educational Research Association
1230 17th Street, NW, Washington, DC 20036
202/223-9485, Fax: 202/775-1824
Contact: Gerald Sroufe, Governmental Professional
Liaison

American Geological Institute
4420 King Street, Alexandria, VA 22302
703/379-2480, Fax: 703/379-7563
Contact: David Applegate, Director, Government
Affairs

American Geophysical Union
2000 Florida Avenue, NW
Washington, DC 20009
202/462-6903, Fax: 202/328-0566
Contact: David Thomas, Manager, Public and
Government Relations

American Institute of Chemical Engineers
1300 Eye Street, NW, Suite 1090 East
Washington, DC 20005
202/962-8693, Fax: 202/962-8699
Contact: Sean Bersell, Director, Legislative and
Regulatory Affairs

American Institute of Aeronautics and Astronautics
370 L'Enfant Promenade, SW
Washington, DC 20024
202/646-7404, Fax: 202/646-7508
Contact: Joanne Padron, Director, Astronomics
Policy

American Institute of Biological Sciences
730 11th Street, NW, Washington, DC 20001
202/628-1500, Fax: 202/628-1509
Contact: Clifford Gabriel, Executive Director

American Institute of Physics
One Physics Ellipse, College Park, MD 20740
301/209-3108, Fax: 301/209-0846
Contact: Richard M. Jones, Senior Liaison, Public
Information

American Mathematical Society
1527 18th Street, NW, Washington, DC 20036
202/588-1100, Fax: 202/588-1853
Contact: Sam Rankin, Director, Washington Office

American Meteorological Society
1701 K Street, NW, Suite 300
Washington, DC 20006
202/466-6070, Fax: 202/466-6073
Contact: Richard Hallgren, Executive Director

American Physical Society
529 14th Street, NW, Suite 1050
Washington, DC 20045
202/662-8700, Fax: 202/667-8711
Contact: Robert Park, Executive Director,
Washington Office

American Podiatric Medical Association
9312 Old Georgetown Road
Bethesda, MD 20814
301/571-9200, Fax: 301/530-2752
Contact: John Carson, Director, Government
Affairs

American Political Science Association
1527 New Hampshire Avenue NW
Washington, DC 20036
202/483-2512, Fax: 202/483-2657
Contact: Catherine E. Rudder, Executive Director

American Psychiatric Association
1400 K Street NW, Washington, DC 20005
202/682-6000, Fax: 202/682-6114
Contact: Melvin Sabshin M.D., Medical Director

American Psychological Association
750 First Street, NE, Washington, DC 20002
202/336-6062, Fax: 202/336-6063
Contact: S. Jefferson McFarland III, Director,
Public Policy

American Psychological Society
1010 Vermont Avenue, NW, Suite 1100
Washington, DC 20005
202/783-2077, Fax: 202/783-2083
Contact: Alan Kraut, Executive Director

American Society for Engineering Education
1818 N Street NW, Suite 600
Washington, DC 20036
202/331-3500, Fax: 202/265-8504
Contact: Ann Leigh Speicher, Manager, Public
Policy and Information

American Society for Horticultural Science
113 South West Street, Alexandria, VA 22314
703/836-4606, Fax: 703/836-2024
Contact: Charles H. Emely, Executive Director

American Society for Microbiology
1325 Massachusetts Avenue, NW
Washington, DC 20005
202/822-9229, Fax: 202/942-9335
Contact: Janet Shoemaker, Director, Public and
Scientific Affairs

American Society of Animal Sciences
9650 Rockville Pike, Bethesda, MD 20814
301/571-1875, Fax: 301/571-1837
Contact: Robert G. Zimbelman, Executive Vice
President

American Society of Human Genetics
9650 Rockville Pike, Bethesda, MD 20814
310/571-1826, Fax: 301/530-7079
Contact: Elaine Strass, Executive Director

American Society of Mechanical Engineers
1828 L Street, NW, Suite 906
Washington, DC 20036-5104
202/785-3756, Fax: 202/429-9417
Contact: Philip Hamilton, Managing Director,
Public Affairs

American Society of Plant Physiologists
15501 Monona Drive, Rockville, MD 20855-2768
301/251-0560, Fax: 301/279-2996
Contact: Kenneth M. Beam, Executive Director

American Veterinary Medical Association
1101 Vermont Avenue, NW, Suite 710
Washington, DC 20005
202/789-0007, Fax: 202/842-4360
Contact: Pamela Abney, Senior Policy Specialist,
Governmental Relations

American Sociological Association
750 1st Street, NE, Washington, DC 20002
202/336-5500, Fax: 202/336-5797
Contact: Marilyn Richmond, Assistant Executive
Director, Government Relations

Association for Women in Science
1522 K Street, NW, Suite 820
Washington, DC 20005
202/408-0742, Fax: 202/408-8321
Contact: Catherine Didion, Executive Director

Biophysical Society
9650 Rockville Pike, Rm 512
Bethesda, MD 20814
301/530-7114, Fax: 301/530-7133
Contact: Emily M. Gray, Executive Director

Computing Research Association
1875 Connecticut Avenue, NW, Suite 718
Washington, DC 20009
202/234-2111, Fax: 202/667-1066
Contact: Fred Weingarten, Executive Director

Conference Board of Mathematical Sciences
1529 18th Street NW, Washington, DC 20036
202/293-1170, Fax: 202/265-2384
Contact: Ronald C. Rosier, Administrative Officer

Consortium of Social Science Associations
1522 K Street, NW, Suite 836
Washington, DC 20005
202/842-3525, Fax: 202/842-2788
Contact: Howard Silver, Executive Director

**Council of Professional Associations on Federal
Statistics**
1429 Duke Street, Suite 402
Alexandria, VA 22314
703/836-0404, Fax: 703/684-2037
Contact: Edward J. Spar, Executive Director

**Federation of American Societies for Experimental
Biology**
9650 Rockville Pike, Bethesda, MD 20814
301/530-7000, Fax: 301/530-7190
Contact: Gar Kaganowich, Director, Office of
Government Liaison

**Federation of Behavioral, Psychological, and
Cognitive Sciences**
750 First Street, Room 5004
Washington, DC 20002-4242
202/336-5920, Fax: 202/336-6158
Contact: David Johnson, Executive Director

Institute of Electrical and Electronics Engineers
1828 L Street, NW, Suite 1202
Washington, DC 20036-5104
202/785-0017, Fax: 202/785-0835
Contact: Chris Brantley, Manager, Government
Activities

Joint Policy Board for Mathematics
1529 18th Street, NW, Washington, DC 20036
202/234-9570, Fax: 202/462-7877
Contact: Lisa Thompson, Congressional Liaison

National Society for Professional Engineers
1420 King Street, Alexandria, VA 22314-2715
703/684-2873, Fax: 703/836-4875
Contact: Bob Reeg, Manager, Congressional and
State Relations

Optical Society of America
2010 Massachusetts Avenue, NW
Washington, DC 20036
202/416-1423, Fax: 202/416-6130
Contact: Susan Reiss, Managing Editor

Society for Neuroscience
11 Dupont Circle, NW, Suite 500
Washington, DC 20036
202/462-6688, Fax: 202/234-9770
Contact: Kelly Mills, Director, Government Affairs

II. Other Organizations

**American Association of State Colleges and
Universities (AASCU)**
One Dupont Circle, NW, Washington, DC 20036
202/293-7070, Fax: 202/296-5819
Contact: Ed Elmendorf, Vice President,
Governmental Relations

AASCU is comprised of chancellors and presidents
of state colleges and universities. It monitors
legislation and regulation and provides a forum for
exchange of information on issues regarding higher
education.

American Forest and Paper Association
1111 19th Street, NW, Suite 700
Washington, DC 20036
202-463-2455, Fax: 202/463-2785
Contact: W. Henson Moore, Acting Vice President,
Government Relations

This association represents wood and paper
companies. It provides information on forest
resources and environmental-related industrial
policies; commissions public opinion research on
resource and environmental issues.

Association of American Universities (AAU)
One Dupont Circle, Suite 730
Washington, DC 20036
202/466-5030, Fax: 202/296-4438
Contact: George Leventhal, Senior Federal
Relations Officer

AAU represents 58 of the nation's top research
universities, as well as four Canadian universities;
works closely with NASULGC (see below).

**Association of American Medical Colleges
(AAMC)**
2450 N Street, NW, Washington, DC 20037
202/828-0400, Fax: 202/828-1125
Contact: Dave Moore, Associate Vice President,
Governmental Relations

AAMC represents medical colleges on research
funding and other issues of interest to the
biomedical community.

**Council of Graduate Schools in the United States
(CGS)**
One Dupont Circle, Suite 430
Washington, DC 20036
202/223-3791, Fax: 202/331-7157
Contact: Thomas Linney, Executive Vice President,
Governmental Relations

CGS is comprised of the deans of the nation's
major graduate schools.

Council on Competitiveness
1401 H Street NW, Suite 650
Washington, DC 20005
202/682-4292, Fax: 202/682-5150
Contact: Mildred Porter, Director, Planning and
Administration

The Council is an organization of leaders of
industry and higher education concerned with the
promotion of policies that will improve U.S.
industrial competitiveness. It publishes a variety of
reports, including an annual "competitiveness
assessment" of the President's proposed budget. A
number of scientific and higher education
associations (including AAAS) are among its
affiliates.

Industrial Research Institute (IRI)
1550 M Street, NW, Washington, DC 20005
202/296-8811, Fax: 202/776-0756
Contact: Margaret Grucza, Director, Research
Services

IRI is comprised of the vice presidents for R&D or
their counterparts in most of the nation's largest
S&T-intensive corporations. The Institute holds
meetings and seminars, conducts studies and issues
publications in areas related to industrial R&D.

National Academy of Sciences (NAS)
2101 Constitution Avenue, NW
Washington, DC 20418
202/334-1513, Fax: 202/334-2419
Contact: Jim Jensen, Director, Congressional and
Governmental Affairs

NAS is a congressionally-chartered, independent
organization that promotes the use and benefits of
science; it advises the federal government on
science and technology issues. The National
Academy of Engineering and the Institute of
Medicine are its affiliates; the National Research
Council is their operating arm.

National Association of State Universities and Land Grant Colleges (NASULGC)
One Dupont Circle, Suite 710
Washington, DC 20036
202/778-0848, Fax: 202/296-6456
Contact: Jerry Roschwalb, Director, Federal Relations-Higher Education

NASULGC represents state universities and land-grant colleges. The Association is active in many areas, including student aid, health, and agriculture and works closely with AAU.

Research!America
1522 King Street, 2nd Floor
Alexandria, VA 22314
703/739-2577, Fax: 703/739-2372
Contact: Mary Woolley, President

Research!America was established to build public support and appreciation for health research. It operates mainly through advertisements and public service announcements in print and electronic media.

The Science Coalition
1317 F Street, NW, Suite 600
Washington, DC 20004
202/662-3716, Fax: 202/628-5035
Contact: Joel Malina, Vice President, The Wexler Group

The Coalition is comprised of 250 universities, companies, and associations from across the country committed to sustaining the federal government's commitment to U.S. world leadership in basic science research.

APPENDIX G

THE CONGRESSIONAL YEAR

The House and Senate follow similar but not identical calendars. Each new Congress convenes on January 3, following the November election, as provided in the Twentieth Amendment to the Constitution. Generally both houses then recess until late in January, when they reconvene for the President's State of the Union Address and budget message. The main exception to this occurs during presidential election years, when Congress must convene in a joint session after the election to count the electoral votes for the President and Vice President.

A considerable amount of informal activity occurs between the election and the initial convening date and during the January recess. These activities include election of leadership, determination of committee and subcommittee assignments, and allocation of office space. After Congress reconvenes in January, legislative work begins (or continues, in even-numbered years). Since congressional committees require some time to prepare legislation, very little legislation is passed in the early weeks of the session.

Congress recesses several times during the year. Most of the dates for recesses vary from year to year; however, they generally follow patterns established by long tradition. Normally, both House and Senate recess for about a week in February around President's Day, in April around Easter, in May for the Memorial Day weekend, in early July around Independence Day, and, often, for the latter part of August through Labor Day. When reading a House calendar it is important to note that recesses are designated as District Work Periods. During Presidential election years, both Houses will also recess for the political conventions.

Generally, Congress aims to complete its work by early October. However, in recent nonelection years, because of increasingly fractious debates over budget issues, Congress has seldom completed business much before Christmas. The date for adjournment is voted on by the House and Senate.

APPENDIX H

LEGISLATIVE BUZZERS, BELLS, AND SIGNAL LIGHTS

All congressional hearing rooms contain signal buzzers or bells and many also have lights. These devices are used to inform Members in hearings of things that are taking place on the floor of the House or Senate. Often these floor activities require Members to leave a hearing abruptly. Few witnesses or observers in these hearings understand the significance of these often distracting signals, which differ between the two houses. Following are lists of House and Senate signals:

House Legislative Electric Bell Signals

1 ring – Teller vote, not recorded.

1 long ring, pause, 3 rings – The start or continuation of a notice quorum call. This call is terminated when 100 Members appear.

1 long ring – Termination of a notice quorum call.

2 rings – Electronically recorded vote.

2 rings, pause, 2 rings – Manual roll call vote. The bells are rung again when the clerk reaches the R's.

2 rings, pause, 5 rings – First vote under Suspension of the Rule or on clustered votes. 2 rings will sound 5 minutes later. The first vote will take 15 minutes with successive votes at intervals of not less than 5 minutes. Each successive vote is signaled by 5 rings.

3 rings – Quorum call, either initially or after a notice quorum has been converted to a regular quorum call. The bells are repeated 5 minutes after the first ring. Members have 15 minutes to be recorded.

3 rings, pause, 3 rings – Manual quorum call. The bells are rung again when the clerk reaches the R's.

3 rings, pause, 5 rings – Quorum call in the Committee of the Whole, which may be immediately followed by a 5–minute recorded vote.

4 rings – Adjournment of the House.

5 rings – Five-minute electronically recorded vote.

6 rings – Recess of the House.

12 rings rung at 2 second intervals – Civil Defense Warning.

Senate Legislative Buzzers and Signal Lights

Pre-session signals: 1 long ring at hour of convening; 1 red light to remain lighted at all times while Senate is in normal session.

1 ring – Yeas and Nays.

2 rings – Quorum Call.

3 rings – Call of Absentees.

4 rings – Adjournment or Recess (end of daily session).

5 rings – Seven and a Half Minutes remaining on Yea or Nay Vote

6 rings – If the lights are turned off immediately, morning business is concluded. If the lights stay on during the period of recess, it means recess during daily session.

MAP OF CAPITOL HILL

N

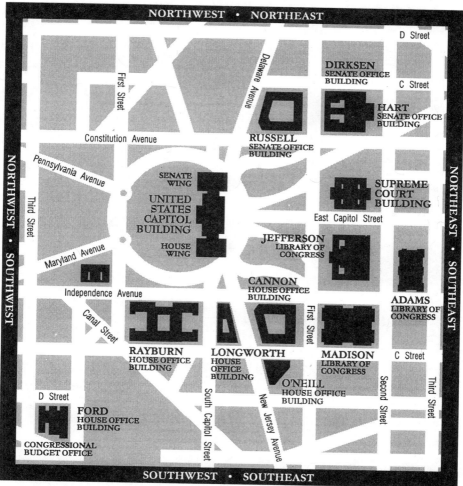

Adapted from *Congressional Yellow Book*. (Washington, DC: Monitor Publishing Company.).

INDEX